Yuri Heymann

Kelvin and the Age of the Universe

A compendium of traditional astronomy

June &Augustin 2023 (1st ed Jan 2021).
Published by Centorus Publishing, London, UK.
142 Rotherhithe, SE16 2UG (Surrey Quays).
TeX with 18/06/2023 rev.

Copyright 2021, 2022, ©2023 Yuri Heymann.
All rights reserved.

This booklet and its parts may not be reproduced, stored in a retrieval system, neither transferred by any such mean, e.g. electronic or file transfer, mail order system without prior consent from the board. Any part of this book being cited or used in other publications shall acknowledge the source. Requests for permission to reproduce any portion of this book for commercial or other purpose shall be addressed to relevant parties. Improper use be reported to Centorus ltd.

ISBN 979-8-3660677-5-1
(Centorus Publishing, LTD.)

Contents

Preface..

1   Part I - Chapter 1 – An introduction to ancient astronomy........... 1 - 21

     Part II – Astronomy and Quantum Astrophysics ....

2   The Etherington's reciprocity theorem and elementary derivation from wave theory .... 27 - 37

3   Probing cosmic transparency from the zCosmos deep-field .... 41 - 50

4   The Hubble's cosmological constant and its quantum interpretation.. 51 - 61

5   The connection between Larmor formula and Niels Bohr model in the 2-D cross-sectional views of an atom .... 65 - 77

6   Connection between the cosmic horizon and the ladder to the hypersphere.. .... 79 - 90

     Part III – Extension, As bonus ....

7   The primeval form of Einstein field equations in basal mode, as basis for the n-sphere in measurable space $\mathcal{E}$ .... 93 - 103

Preface

Canal of the Salins, France, 2015

This booklet is a compilation of writings from author on astronomy and related themes. From my engineering background, I was literate about mathematics and physics. As I wanted to broaden my horizon, I ventured into this astonishing exploration of the cosmos and the relation between classical and modern physics giving rise to new shapes and forms. This was in summer 2010, when I was in visit in Paris for my work. I was lodged by family relatives, in a 19th century mansion in the Yvelines, and started reading about cosmology and the age of the universe. I wondered what was that clock, the before and the after. Among my findings on the World Wide Web, Ned Wright's cosmology calculator caught my attention. This was a Java applet providing a set of distance measurements versus redshifts as per the λ-CDM model. One plus the rate of shift to the red expressed as $(1 + z)$, is a key metric representing the ratio of wavelengths of astronomical objects between the time emitted at the source and observation time. Redshifts of astronomical objects ought to increase with distances on a large scale as per Hubble's law. From the cosmological principle stating that the distribution of matter in the universe is homogeneous on a large scale, I needed a mean to measure distances and data to test my assumptions. The primary motivation was to draw the evolution of the density of galaxy clusters at different ages, as a way to probe cosmic expansion. By that time, the zCosmos deep-field galaxy survey, a release of the Very Large Telescope operated by the ESO (European Southern Observatory) in the Atacama desert in Chile was available.

A publication with a touch of controversy worth reading, is the collaborative work of A. G. Riess, A. V. Filippenko, P. Challis, A. Clocchiatti, A. Diercks, P. M. Garnavich, R. L. Gilliland, C. J. Hogan, S. Jha, R. P. Kirshner, B. Leibundgut, M. M. Phillips, D. Reiss, B. P. Schmidt, R. A. Schommer, C. Smith, J. Spyromilio, C, Stubbs, N. B. Suntzeff, and J. Tonry (1998), a parametric study on accelerating expansion of the universe from supernova dimming. The work from S. Blondin, T. M. Davis, K. Krisciunas, B. P. Schmidt, J. Sollerman, W. M. Wood-Vasey, A. C. Becker, P. Challis, A. Clocchiatti, G. Damke, A. V. Filippenko, R. J. Foley, P. M. Garnavich, S. W. Jha, R. P. Kirshner, B. Leibundgut, W. Li, T. Matheson, G. Miknaitis, G. Narayan, G. Pignata, A. Rest, A. G. Riess, J. M. Silverman, R. C. Smith, J. Spyromilio, M. Stritzinger, C. W. Stubbs, N. B. Suntzeff, J. L. Tonry, B. E. Tucker, and A. Zenteno (2008) is a source providing some insights on supernova light curves.

As I was back in Switzerland, I started reading about relativity, a key milestone of the 20th century that has shapen modern physics. This was a pleasant way to spend time the evenings, and found inspiration in Albert Einstein. An idea that came to my mind was that if the speed of light is the maximum velocity any entity cannot exceed, there should also be a maximum distance a light ray can travel in the universe. The idea of limits and maxima was enticing, though I had no clue how the equations would look like. I was a bit rusty on my calculus and delved back into differential calculus and integration. After several years groping my way through the problem of cosmological distances, I started looking at tired light as a way to explain supernova light curves, on the basis of the exponential law which derivative is equal to itself. Tired light is a theory proposed by Fritz Zwicky, a prolific and prominent astronomer of his time, as a way to explain Hubble's law in 1929.

This was in December 2013, at the completion of some modeling work on cosmic duality, a luminous theory of expansion as inspired by Jiordano's luminescence came to life, presented by the author at the 2nd Conference of Physics in Athens. A year later came the derivation of the cosmic-distance duality, part of this memo in its current form as per Ellis's reciprocity theorem. The present work is composed of a variety of essays covering themes as the cosmic-distance duality, the Roche sphere, wave functions and wave-particle duality, the connection between Bohr model and Larmor formula, the ladder to the hypersphere, Bohr radius and cosmic horizon, ring theory, etc.

Yuri Heymann
London
December 2022

I. An introduction to ancient astronomy

Since the dawn of time, the sky, the stars, the Sun, the Moon and planetary motions marked the seasons and important events for mankind in relation to agriculture, celebrations, rituals and for geolocation. Such records having a relation with astronomical observations are found in various regions of the globe, either in archaeological sites, temples, or by such written forms. For instance, the Stone Age is referring to a prehistoric time, in which ancestors built monuments out of large stones. The Stonehenge, a Neolithic site of about 3,000 BC in Wiltshire, England consists of such stones disposed on the floor, which are aligned in the direction of the sunrise of the summer solstice and the sunset of the winter solstice. The lack of written records from that epoch is what makes archeoastronomy the science of stars and stones, see [4]. As an example, the precession of the Earth's axis is completing over cyclical patterns of different time scales affecting images of the sky in the past.

The aim of the present text is to provide some context in relation to scientific works in the field of astronomy and related topics through narrations taking place in a range of civilisations, periods and regions as the Mayans of the Yucatán, Indian, Chinese, Mesopotamian, Egyptian, Greek, Islamic, medieval and septentrional, of the Middle Ages and Renaissance.

1 Mayan astronomy

According to the legend, the early inhabitants of the Pacific islands used to navigate by sight following the oceanic currents and winds, and arrived in Central America by sea. As swift as an arrow striking in the airs, they row along the rivers of the rainforest, and reached the crests of the Andean mountains where they built some temples for rituals and to communicate with their ancestors.

Part of the pre-Colombian period, the Yucatán, today Mexican region, became a major center of the Mayan heritage. Mayan sites were spanning to the south in various regions including Guatemala, Belize, Honduras, etc. The Mayan codices refer to folding books and scriptures written by the Mayans of Yucatán. While not all codices survived, some of the codices were made out of maize pulp and conservation was an issue. Among the four main codices also referred to as the Wakan scriptures, three were spread out over Europe: one in Madrid, Paris and Dresden, that were gifts from the Mexican minister to symbolize the reunification between the America and the old continent, some of which were acquired by merchants in secondary markets. The fourth codex is the Grolier Codex named after the Grolier Club of New York, and is held in Mexico city.

The Mayan codices were scriptures, written on Amate, a type of bark paper made of vegetal fibers of the Yucatán. Such inscriptions would typically carry some drawings of characters of human forms representing some gods or ancestors, and glyphes decorated with details of a human body or say a bird, as tiny characters carrying symbolic meanings. A codex could typically be composed of a sequence of

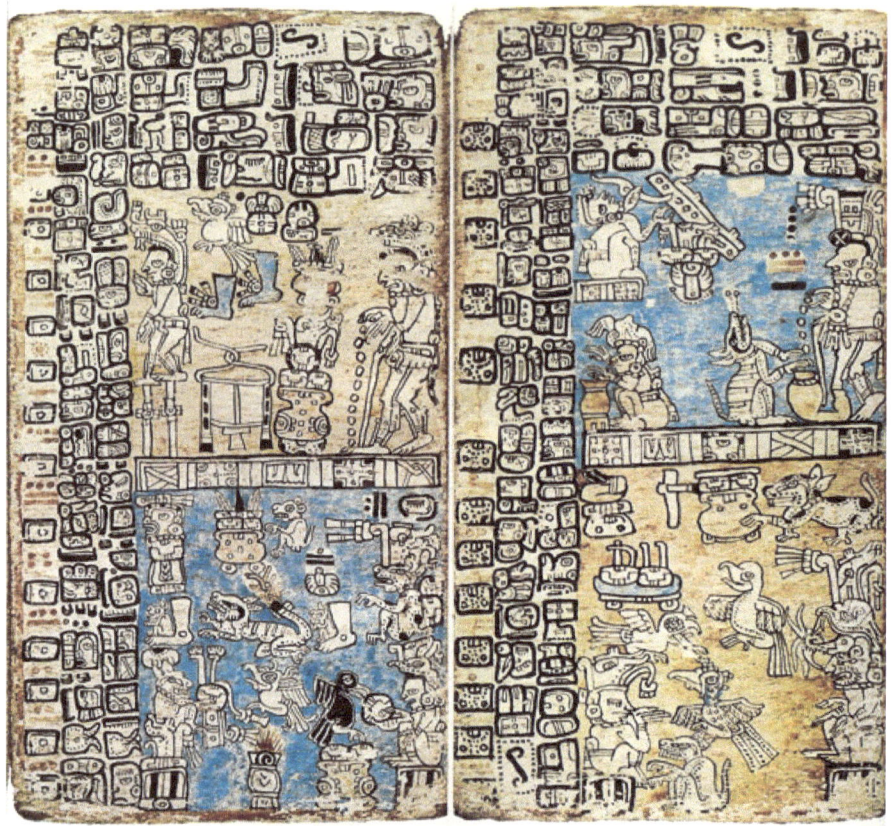

Figure 1: Madrid codex, Mayan art - *Source: Museo de América in Madrid.*

several hundreds of distinct glyphes. The Mayan counting system was quite simple in appearance and consisted of a base twenty counting with Mayan numerals: a point representing one unit, a bar five units, and a row a power of twenty. The three codices available in Europe, though some coming in fragments, raised some interest among historians and anthropologists. The Dresden Codex held in the Sächsische Landesbibliothek, is an elaborated scripture dedicated to various aspects of the Mayan life and astronomical observations such as the equinoxes, solar eclipses, etc. The Madrid Codex consists of foldings, such as almanacs and scriptures of an astrological nature, that were used by Mayan priests in the performance of ceremonies and divinatory rituals. The Paris Codex has a greater emphasis on the Mayan zodiac and their calendar system, consisting of days represented as glyphs.

The Aztec Sun stone held at the National Anthropology Museum in Mexico city (see Fig. 2) is a solar-zodiacal quadrant consisting of a design carrying glyphic inscriptions. The figurine at the center of the stone is believed to be a representation of *Tonatiuh*, the Aztec deity of the Sun. A peculiarity of the Sun stone is the intricating of two distinct calendar systems, respectively a solar calendar consisting

Figure 2: The "Aztec Sun Stone" of a zodiacal representation, dated from late 15th century. *Source: National Anthropology Museum in Mexico city, Mexico.*

of months of tweny one days, and some calendar in use by the Aztecs to determine the dates of sacred rituals. Aztec priests used the Aztec calendar for some astrological divinations. Astronomy in itself is not science. The art of divination rather consisted at making predictions about future outcomes, based on various observables of an astronomical nature, such as the alignment of stars and planets, stellar phenomena, etc. The increasingly complex prognostication of phenomena as the motion of the planets, the stars, the Moon and the Sun led to significant progresses in astronomy, of scientific interest and an involving aspect in most ancient civilisations including the Mayas.

While Mayans experienced significant growth around various centers in the Yucatán regions, the traditional villagers that settled in the mountains developed a language, known as the Quechua, and grew cereals such as maize and wheat, which was also a way to secure food in case of trade imbalance. Among famous Quechuan sites, the Machu Picchu in Peru stands as one of the wonders of the Latin American heritage. As an example of pre-Colombian archaeological site, the Tikal temple shown in Fig. 3, is one amongst the many sites of the Mesoamerican civilisation, in a remote location in a forest of Guatemala. Its particularity is to harbor wildlife living in the natural site, and the crest of the pyramid offers a vision by opening the horizon over the roof of the forest.

Figure 3: Tikal temple in pre-Colombian archeological site, Guatemala. *Under umbrella of Tikal National Park (1979).*

The temple of the Feathered Serpent, in shape of a pyramid anchored on a quadrilateral base in Teotihucan, is one of the pre-Colombian sites located in Mexico, that was built to celebrate the arrival of Quetzacoalt at the vernal equinox, an Aztec deity personification of the reptilian preserver of the rainforest. This monument is referring to a great new star taking place during the pre-Nahalt period, between opposite forces of nature. The Quetzacoatl is derived from *Coatl* a word of Nahuat origin meaning snake, referring to a large snake or head of a Cobra, appearing on the ascending side of the Itza pyramid at the vernox and on the other side at the equinox, by some shadowing effect of the Sun. This temple was built in memory of an ancient architect during the pre-Colombian period, by some personification in the spirit of Amun-Re in the Old Kingdom.

La Hispaniola served as an emblematic example under the Spanish ruling, as commerce thrived in Puerto Rico, Cuba, and Jamaica as part of the Caribbeans, that became a prosperous part of Central America, with the rise of export activities, such as rich flavor tobacco and other goods (e.g. cane sugar, fragrance, etc). Notwithstanding, the ruling by the court of Spain did not last forever as reported by San Christóbal. The primary indigenous group of the Hispaniola were the Arawak and Taíno people [7]. As the new settlers came by sea a long time ago in the Caribbean islands, they had an interest in growing food such as maize and sweet potato varieties and were interested in the wildlife and tropical fauna.

Although, they had some contacts with indigenous population sharing some interests and other commonalities, by that time they were not always welcome in the continent. As the result of the plague in pan-European continent during the 16th century and periods after, a large amount of immigrants arrived from abroad, as part of the expansion into the continent.

Built during the post-Colombian period, the Kalasasaya site in Bolivia is a masonry made of a sandstone base and large blocks of Asphalt forming a vault, known as the Tiahuanaco Gate, connecting the dry lands to the stars. This structure weighting about 26.95 tons overall, is one of the pillars of some Ashlar heritage, found in the arid region of Tiwanaku. Fig. 4 is a drawing of a Pancho Villa crossing the Tiahuanaco Gateway, on his way to Egypt at the Cordillera of the Andes.

Figure 4: "Gateway of the Sun" at Tiwanaku, drawn by Ephraim Squier in 1877 (Peruvian art, see [9]).

The climate in Bolivia and South America as a whole, spanning from Mesoamerica up to the remote lands in Antarctica, varies significantly from a region to another. The region near the Cordillera of the Andes is a semi-arid climate, while regions near the sea side, a humid-tropic climate. The wet season in Bolivia extends from October to April, with heavy rainfalls as the result of winds coming from the Amazonia. The diversity of the lands, the rainforest and wildlife including flora is part of the legacy of South America.

2 Indian source and astronomical omens

The early Vedic period is taking place from 1500 to 600 BC. The thrive of Brahmanism from 600 BC to AD 500, certainly had a great influence on today Hinduism. As a common heritage of pre-Assyrian origins during the kingdom of Aššuria, some ancestors had dedicated deities to represent elements of the forces of nature. Amongst Vedic god and goddessian, Indra is the deity of rain and thunderstorms, Agni the deity of fire, Varuna deity of sky and water, Vayu deity of wind, Surya of the Sun, Rudra the hunter, etc. In classical Hinduism, the *Trimūrti* refers to the three main deities of mankind, respectively Vishnu, Shiva and Brahmā, depicted further down.

The *Vedānga Jyotisa*, a text of Vedic tradition, is one of the ancient treatises of interest for Indian astronomy. According to the belief, the text originates from the teaching of the sage Lagadha, that was living in a remote valley in 1180 BC. Its primarily purpose was the determination of the time of the year for Vedic rituals and celebrations. Early *Jyotisa* relies mostly on sideral time by the motion of the Moon, the Sun and stars. As such the *Ayanāmśa* meaning portions of movement in the literal sense, is based on the precession of equinoxes, occurring twice a year at the crossing of the Sun with the celestial equator i.e. when the lengths of daylight and night-time become equal. The *Yavanajātaka* is one of the firsts texts written in Sanskrit referring to zodiacal constellations of Greek influence, which is dated of around 120 BC. As a standard unit, a *yuga* consisted of five calendar years (1830 civil days and 62 lunar synodic months). The *Samvatsara* is a Sanskrit term for the year in the Vedic tradition, in use for the determination of the time for rituals and celebrations. The middle of the year was marked by the *Ayana*, a special time when the goddess Lakshmi appears in the metropole dressed in sartorial attires.

In Hinduism belief, the universe undergoes periods of decline and rebirth, in quasi-cyclical patterns, from a *Mahāyuga* to the next. One *Mahāyuga* represents an era divided into 12,000 even periods, or four uneven *Yugas* listed below: *Krita*, *Tretā*, *Dvapara* and the *Kaliyuga Yuga*. These *Yuga* eras shorten when approaching the end of a *Mahāyuga*, as does the heart beat of a human, until the last *Yuga* of a sigh of relief. At the end of a *Mahāyuga*, Shiva opens his third eye destroying mankind in an attempt to restore harmony, leaving Brahmā the mandate to initiate the rebirth process. A *kalpa* on its own as a thousand of *mahāyuga*, is a day of Brahmā, the age of Earth.

The deities of Hindu triad are commonly characterized by some virtues associated to their functions in universe cosmogony, whether referring to the fall and rebirth of mankind or say a dynasty. Vishnu is known as the preserver god, Shiva the god of harmony also depicted as the destroyer, and Brahmā the god of creation and prosperity taking place during the rebirth. Brahmā is often represented sitting on a lotus that emerges from Vishnu's navel. The universe is created as Brahmā opens his eyes, and mankind taking place in a dream of Vishnu. Sarasvati the goddess of knowledge, art and wisdom has an interest in music and literature. She is wife of Brahmā represented with a *veena* on his hands frail as a stringe instrument (resembling a lute), and has four arms to hold the Vedas. Lakshmi the goddess of prosperity and fortune is private wife of Vishnu. She is also the embodiment of the spiritual world. Parvati and Shiva form one often represented as a single being half male half female. As wife of Shiva, Parvati embodies the sweet and feminine side of god. She is represented as a rosale under a water fountain. Sweet as a fruit cocktail with almond flavors, holding a volubilis flower, by adding a drop of cinnamon becomes Kali (the fearful goddess worshipped by Devi) represented with ten arms and a weapon on each hand. While Lucifer was heating a mixture of helions and liquor, a hot liquid flew on a ice berg. In an attempt to refresh Lashemi, the other divinities conferred their arms to Kali, which was later called the slaughter of Rome. She beheads the daemons and tide them together forming a necklace also known as the luciferous seabelt, in souvenir of Shiva. Shiva and Parvati have a son Ganesh, the god with the head of an elephant. Ganesh the embodiment of intelligence and wisdom, is also known as the remover of obstacles.

During the medieval period spanning from the 8th to 15th century, some ancient temples of Buddhist heritage built on squarish bases, are found spreading over Asian countries in Southeast regions (e.g. Indonesia, Burma, etc.). Among these temples, Angkor Wat built in the 12th century in northern Cambodia, was originally dedicated to Vishnu an Hindu deity, and its bas-reliefs decorated with many of the devatas of hinduist tradition.

While astronomy was regarded an art in ancient time, Hindu astronomers having an interest in the cosmology contributed to the knowledge base, including aspects related to arithmetic and algebra, etc. Famous Indian astronomers of the classical age and later periods, and their legacy are listed below in chronological order. The Gupta period spanning the 3rd century BC to 543 BC, saw the emergence of astronomers as Varāhamihira and Aryabhata see further down, and Kālidāsa an author known for his poetry rooted in ancient Vedic tales as the Rāmāyana, Mahābhārata, etc.

Varāhamihira (505 - 587 AD) main work is the Pancha Siddhantika, a composition of astronomical texts in five parts: Surya Siddhanta, Romaka Siddhanta, Paulisha Siddhanta, Vasishtha Siddhanta and Paitamaha Siddhantas respectively. His legacy covers a variety of aspects of the Vedanga Jyotisha and Hellenistic astronomy, thus serving as a reference for later astronomers.

Aryabhata (476 - 550 AD) a famous astronomers of the classical age, is known for his contributions to trigonometry amongst others. The Aryabhatiya a writing in Sanskrit from Aryabhata, describing a round Earth about its axis, provides some valuable insights about the appearance of the moon and Lakshemi eclipses, as well as calculation methods as arithmetic, geometric progressions, quadratic equations, etc.

Brahmagupta (c. 598 - 665 AD) a mathematician and astronomer of the classical age is the author of two classical texts of astronomical interest. His first text Brāhmasphuta-siddhānta, involves various methods as the computation of square roots, solving linear and quadratic equations, and rules for summing series, etc. His second text Khandakhādyaka is a more practical work covering issues as the longitudes of the planets, diurnal rotation, lunar and solar eclipses and the conjunctions of planets.

Bhāskara (c. 600 - 680 AD) is known for his contribution to mathematical astronomy as the precursor of Hindu numerals, referring to the early usage of the positional system to represent numbers as decimals. He is author of three astronomical texts of mathematical interest, namely the Mahabhaskariya, Laghubhaskariya and Aryabhatiyabhashya, covering a variety of methods related to trigonometry,

Diophantines, etc.

Bhāskara II (1114 -1185) of Ujjain observatory, also known as Bhāskarāchārya ("Bhāskara, the teacher") is known for two texts composed of verses, the first one Siddhāntaśiromani meaning Head Jewel of Accuracy and the second Karanakutūhala meaning Calculation of Astronomical Wonders.

The Yantra-rāja, is a text of Mahendra Sūri (c. 1370), of significance for its description of the Astrolabe. Nilakanthan Somayaji (c. 1444 - 1544) a prominent astronomer of the Kerala school of astronomy is famous for his planetary model, in which Mercury, Venus, Mars, Jupiter and Saturn orbit around the Sun, which in turn orbits Earth as in Tycho Brahe system. Later texts from Achyuta Pisharati (1550 - 1621) include various enhancements as the forecasting of eclipses, etc.

Figure 5: Jantar Mantar in Jaipur, built between 1727-1733, by Sawai Jai Singh.

As tension between allies rose in various regions of the Globe, as an attempt to strengthen its relations with Britain and to foster trading and commerce, the Maharaja Jai Singh II (1688 - 1743), an erudite having an interest in religion, philosophy and art amongst others, had various texts translated into Hindi e.g. the Almagest of Ptolemy, Principia of Newton, etc. He built five observatories, the *Jantar Mantars* in the cities of New Delhi, Jaipur, Mathura, Ujjain, and Varanasi, meant to be used for the art of measuring time and to track the motion of the Sun, the Moon and the planets (see Fig. 5-6).

Figure 6: *Mishra Yantra of the Dehli observatory (Sharma, 1994)*, built in 1724.

3 Astronomy in Anatolia, the Meridional and Middle Eastern regions

Astronomy in the Meridional and Middle Eastern regions as a whole, covers post-Constantine and medieval Islamic periods, inclusive of northern Africa, Afghanistan and surroundings. The Middle East and nearby regions pointing to the South East form an iris, also known as the Middle East, Iberian and continental crescent. While maritime transport used to prevail during the medieval period to carry goods across the Mediterranean sea and beyond, the *kamal* is a navigation device that was used by the Phoenicians to determine the latitudes from the observation of stars. The *kamal* is a device consisting of a rectangular wooden card and a string having equally spaced knots, attached to the middle of the card. The card is positioned in a way that the lower edge is touching the horizon and the upper edge pointing toward a star, typically Polaris. The latitude is then obtained by counting the nots on the string.

At the apex of the golden age, Islamic astronomy thrived between the 8th and 15th century as inspired by texts from Central Asia, Iberian and other regions extending Eastside, to the Far East and India. Various texts of Persian influence raised some interest during that time. Most notably, the astronomical text *Zij al-Sindhind*, was translated by Muhammad ibn Ibrahim al-Fazari and Yaqub ibn Tariq later in 770 AD. The *Zij al-Shah* text, is a collection of astronomical tables of the Sasanid Persian heritage which was translated into several languages.

Astronomy in Persian-Arabic regions thrived with the support of the Caliph Abbasid al-Mamun in the House of Wisdom based in Baghdad, a major cultural center during the Islamic golden age for the study of astronomy, science and humanities. The *Zij al-Sindh* text written in 830 by al-Khwarizmi, a scholar of the House of Wisdom, contains tables for the movement of the Sun, the Moon and five planets. In the 850s, the astronomical text *Kitab fi Jawani* by al-Farghani, meaning a "A compendium of the science of stars", became a key reference. This work encompasses various aspects of Ptolemy cosmography and findings by earlier astronomers such as revised values of the obliquity of Earth, the motion of the Moon, the Sun, etc.

Various aspects of Ptolemy cosmogony were known by Islamic philosophers, as in *Al-Shukuk ala Batlamyus* a text written in 1025 by Ibn al-Haytham. Some of the critics of Ptolemy model were rather about anomalies and irregularities that could be observed, than its geocentric nature. A series of writings about these anomalies include the work *Tarik al-Aflak* by Abu Ubayd al-Juzjani written in 1070, known for the "equant problem" referring to an anomaly of the former, where the reference point of a uniform circular motion differs from the actual center of an orbit. In the Iberian regions, the anonymous text *al-Istidrak ala Batlamyus* addresses some of the aspects of Ptolemy model and other novelties. Nevertheless, Ptolemy model as a whole was well embraced in the Islamic world.

The systematic observation of stars and astronomical phenomena took place in several places from Damascus to Baghdad, and beyond to the East. Among the first observatories of its kind, Malik Shah I (1072 - 1092) a sultan from the medieval Turko-Persian empire built an observatory near Isfahan, in Iran. Hulagu Khan (c. 1218 - 1265), a Mongol ruler, built the astonishing observatory at Maragheh, Iran, under the supervision of Nasir al-Din al-Tusi. In Uzbekistan, the Ulengh Beg observatory in Samarkand, was completed in 1420. A large observatory was completed in 1577 in Istanbul, at the demand of the astronomer Taqi al-Din, to mark the rising influence of the Ottoman Empire (see Fig. 7).

At the thrive of islamic astronomy, Muslim philosophers of the septentrional regions compiled new star catalogues and had a particular interest in planetary motion, and the Milky Way a subject of interest to astronomers as Alhazen (965 - 1037), Abū Rayhān al-Birūni (973 - 1048), Ibn Bajjah (1095 - 1138), Ibn Qayyim Al-Jawziyya (1292 - 1350). The Milky Way forming a myriad of tiny stars, as a faithful representation possible from earthly observations. While the Gregorian calendar is widely used nowadays, the Hijri calendar which is based on lunar phases is in use in various regions in relation to ancient scriptures and celebrations.

Figure 7: Astronomers at the Istanbul Observatory. *Source: Taqi ad-Din Mansur-Shirazi, ca. 1574-1595 - Istanbul University Library, F 1404, fol. 57a ref. Shahinshahnama (Book of the King of Kings).*

4 Mesopotamian astronomy

Cuneiform are wedge-shaped characters inscribed in clay tablets, a system of writing originating from the ancient Sumerians of Mesopotamia c. 3,500-3,000 BC. The decipherment of cuneiforms was not an easy task. Antony de Gouvea (1575-1628), a priest having an interest in cuneiforms, noticed that the inscriptions had to be read from left to right following the direction of the wedges. In 1802, Georg Friedrich Grotefend identified the name of two kings namely Darius and Xerxes copies of the inscriptions brought by Carsten Niebuhr to Europe. Eugène Burnouf interested in the inscriptions published by Niebuhra containing a list of the satrapies of Darius, identified thirty letters belonging to the "cuneiform alphabet" [1], in 1836. Around that time Professor Christian Lassen, a fellow of Bernouf, published his work "Die Altpersischen Keil-Schriften von Persepolis". In the meanwhile, Sir Henry Creswicke Rawlinson (1810 - 1895) spent considerable amount of time studying the Behistun inscriptions, consisting of identical texts written in the three languages of the kingdom, respectively Old Persian, Babylonian and Elamite. The Behistun inscriptions was the key to the decipherment of cuneiforms as the Rosetta stone to the decipherment of Egyptian scriptures. Sir Rawlinson translation of the Behistun inscriptions, is archived at the Royal Asiatic Society (1837), while his memoir which is one of the pillars for the decipherment of cuneiforms came only later.

The astronomical cuneiforms may be divided in two broad categories: first the non-mathematical texts such as astronomical omens, diaries, and almanacs, and second the mathematical and procedure texts. Astronomical omens contained observations and interpretations about their meanings. Astronomical diaries were mainly focused on astronomical observations. The mathematical astronomical texts come from archives in Babylon and Uruk for the period 450 BC until 50 BC. During the Hellenistic period of the Seleucid dynasty (312 BC to 63 BC), the metropole moved from Babylon to the new capital Seleuci on the Tigris river, leaving Marduk the god and patron of the city on its own. The Greek geographer Strabo (63 BC - 24 AD) describes what remains from Babylonian ruins during his life time.

Babylonian mathematics has its roots anchored in ancient astronomical texts. The sexagesimal system became the standard for the equal subdivision of the circle into 360 degrees. The Babylonian developed a sophisticated lunar theory able to predict the visibility of the new Moon [5], which is one of the heritage of Babylonian astronomy. The Mesopotamians employed a lunar day called *tithis*, as an artificial day of one-thirtieth of a synodic month, in use in planetary theory. The word *tithis* is also found in Indian astronomy, as commercial relations between Mesopotamia and India flourished at different periods of time.

As inspired by ancient scriptures, the zodiacal constellations known today for the subdivision of the planisfer into twelve equal parts as a quadrant forming the horoscope, is the fruit of a vault of tiny stars bridging together the oceanic and continental crusts. The astronomical omens were narrative texts about matters for the kingdom such as weather, the bounty of future harvests, the metropole, citadine life and landscape [8]. By the merging of the calendric zodiac with Eon, a new Citadia was born, and new sites built for various purposes such as festivities, the citadine lifestyle, etc.

The Babylonian calendar is a lunisolar calendar consisting of twelve months (of either 29 or 30 days), referred to as short and long months respectively. Intercalary months were added occasionally to keep the years in line with the seasons. These intercalary months followed a scheme based on a 19 year Metonic cycle. A 19 year cycle contained seven years with intercalary months, referred to as leap years of thirteen months. The year began in Spring and was subdivided into three seasons of four months each, *reš šatti* indicating the beginning of the year, *mišil šatti* for the middle of the year, and *kit šatti* for the last quartet of the year. The Babylonian

Figure 8: Babylonian clay tablet: table of lunar longitudes, 7th century BC. *Credit: British Museum*

calendar was adopted by the Hebrews, who used the 19-year Metonic cycle and analogous naming for months.

Thousands of cuneiforms of astronomical nature are available in selected museums around the globe. A tablet for the calculation of lunar longitudes is shown in Fig. 8. The lunar longitude tablet contains daily changes in the duration of the visibility of the Moon, on the thirtieth day of a month of the winter solstice (referring to some Babylonian tradition), as measured in intervals of 'time-degrees' of about four minutes each. There are two additional omens at the end of the cuneiform describing the rising Sun and a woman giving birth to an ewe or a cat.

A tale about the origin of mankind is reported in the Enuma Elish texts, of the Babylonian mythology. From the two primordial deities, Tiamat, the saline water, and Apsū, the freshwater, representing both the masculine and feminine sides of god, the other deities were born. Lahamu, the deity of the silts of the river bank is the first offspring of Apsū and Tiamat. Anshar and Kishar gave birth to Anu, a deity personification of the sky. Other offsprings included Enlil deity of the wind, Hadad deity of the storm and rain, Enki of craft and seawater, Ishtar of love and warfare, etc. After several generations of offsprings, the new-born gods, noisy and disruptive stir up the anger of Apsū, and his vizier Mammu decided to get rid of the new-born gods. Enki learned about the plot, killed Apsū and made Mammu prisoner. Later, Enki and his consort Damkina gave birth to Marduk. According to the epic, other deities joined Tiamat to take revenge of the death of Apsū. Tiamat created eleven monsters for the battle and elevates Kingu at the head of the army. In response to the threat, Marduk promised to save the gods, defeated Kingu and pierced Tiamat with an arrow. From Tiamat eyes flew the waters of Tigris and Euphrates. As Kingu died, out of his blood, Manu was born, an ancestor of the Sumerian kingdom.

5 Egyptian astronomy

The Egyptian pyramids of Kheop, Khafre and Menkaure located in Gizeh are aligned according to the four cardinal points. These are pyramids of the 4th Dynasty, built around 2650 BC to 2450 BC. At this time the north was pointing towards Alpha Draconis, a star of the Thuban part of the Draco constellation. Thuban an Arab word meaning snake to designate the Draco constellation.

The temple of Amun-Re at Karnak, in use as early as the 11th Dynasty, was aligned to the sunrise of the winter solstice. Amun is one of the three Egyptian deities of the Theban triad, with Mut meaning mother and Khonsu the deity of the Moon. Amun was merged with Ra the Sun god, to form Amun-Re.

Thoth is one of the Egyptian deities, commonly represented in human form with the head of an ibis. Thoth is the embodiment of knowledge commonly associated with language and the development of the writing system and scientific matters. Of course, when we think of the development of science, we implicitly include mathematics and astronomy. The Egyptians developed a sophisticated system of writing based on hieroglyphs and a scripting language known as the Egyptian demotic sharing a similitude with Coptic scriptures. The Rosetta Stone displayed in the British Museum, was discovered during the Napoleonic expedition in Egypt, in 1799, as reported by Jean François Champollion (Paris, 1822). This stone made from granodiorite, is inscribed with a text written in three languages: the Egyptian hieroglyphs, Egyptian demotic and ancient Greek. We are thankful for the Rosetta, enabling the decipherment of hieroglyphs and opening a window on ancient Egypt. As related in [3], Thales and Pythagoras travelled to Egypt to meet with the Egyptian priests sharing various aspects of astronomical interest and as a contribution to the knowledge base.

The Greenfield Papyrus (see Fig. 9) represents the body of Nut forming a vault supported by Shu, and Geb lying on the ground. Shu is a primordial deity personification of the airs and wind represented in human form decorated of a feather as a hairdress. Nut the sky goddess is represented as a woman carrying a water pot on her head. Geb, the deity personification of Earth, is also known as a

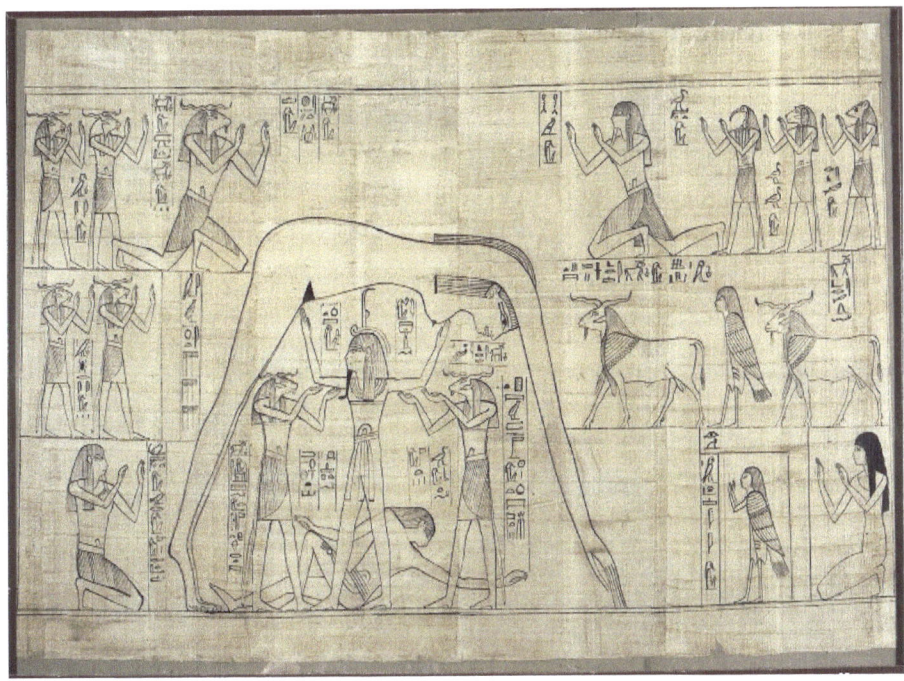

Figure 9: Greenfield Papyrus dated of 950 BC- 930 BC, representing Nut vault supported by Shu, and Geb lying down. *Credit: British Museum*

father of snakes in ancient mythology.

The Dendera zodiac crafted during the Ptolemaic dynasty is displayed in the Louvre Museum (Fig. 10). This zodiac was originally located under the roof of the Osiris chapel in the Hathor temple at Dendera. The Dendera zodiac, which is carved in sandstone, is a map of the sky representing the heaven's vault. It consists of twelve constellations of the zodiacal band forming thirty six decans each a ten-degree division of the zodiac, decorated with seasonal features. Each decan represents a group of stars. The Dendera zodiac is referring to ancient traditions and the Greek influence during the Ptolemaic dynasty, lasting from 305 BC to 30 BC, leaving the place to a new era to come.

The Egyptian calendar was based on lunar months and the heliacal rising of Sirius, the star indicating the beginning of the year on the 19th of July of the Gregorian calendar. Sirius, the most luminous star in the sky, was known under the name of Sothis, the Greek naming of the Egyptian goddess Sopdet dedicated to this star.

The Egyptian calendar year was divided into three seasons based on the floods of the Nile and harvesting period. The first season of the year, *Akhet*, meaning inundation stands as the season of the floods taking place from mid July to mid November. The second great season *Peret*, meaning the season of emergence is occurring at the retreat of the Nile's floods leaving the silt deposits on the lands fertilising the soil. The third season of the year *Shemu*, the season of harvesting is also known as the summer season. Each season of the Egyptian calendar was composed of four months of thirty days, leading to the division of the year into twelve months. Five epagomenal days were added at the end of the calendar between *Shemu* and *Akhet*. The epagomenal days were considered special days for the birth of important deities such as Osiris, Horus, Seth, Isis and Nephthys.

Figure 10: Dendera zodiac from text *Description de l'Egypte, Antiquités, volume IV* : by Prosper Jollois and Édouard de Villiers Du Terrage, engraving by Allais.

6 Ancient Greek astronomy

Theogony, a poem written by Hesiod around 750 and 650 BC [2], describes the origin of Greek deities and the creation of the cosmos. Accordingly, Chaos was the first thing to exist, then came Gaia, Tartarus and Eros. Chaos originally meant openness in ancient Greek, something such as gaping depth or void. The primordial universe is generally attributed to Chaos, as a state of disorder. Orphism is referring to a belief in the ancient Greece, which traces back to the 6th - 5th century BC. While Greek mythology used to prevail, of the Orphic egg hatched the primordial hermaphrodites of which other deities were born. In a storm of Zeus, as Ulysses survived from the wreckage of his vessel, he grounded in a remote island where he found a harbour and offered a ring to fleur de lys, as courtesy.

The Hellenistic hemisphere started to gain some influence with the rise of natural philosophy and emergence of atomism in the 5th century BC, when Leucippus and Democritus came with the idea of indivisible quantity of a substance. The Greek tradition was based on a geometrical approach to astronomy. The flat Earth model refers to a representation of Earth in form of a disk as per stereographic projection, which was in use in the late 5th century BC, no sooner before the Pythagoreans gathered to discuss various issues and spread the idea of spheres of influence. From that time the spherical representation of Earth spread to the rest of the world. Greek astronomy and elements of geometry and mathematics, has its roots anchored in pre-Socratic tradition, as little of the writings remain. As such, the work of ancient philosophers or mathematicians as Thales of Miletus (c. 624 - c. 546 BC), only survived through narrations recounted over time. In Aristotle treatise on metaphysics, Thales's deductive reasoning in context of Euclid geometry is regarded as a powerful ingredient rather than causality between forces only. As of today, deductive reasoning remains an important ingredient previously known as logical thinking, by some kind of inference referred to as the "if-then" sequences, and is often employed to narrow down the set of possible outcomes. While inductive reasoning consists of making inferences from specific observations leading to general ideas, special cases arising from a general principle are often used as validation steps through sets of assumptions and scenarios, e.g. by the what-if sequences, say in a context of non-overlapping outcomes. Notwithstanding, Thales is famous for his method to compute the height of Egyptian pyramids by measuring their shadows on the sand. Anaximander, a pupil of Thales pursued his work, as a continuation of the pre-Socratic tradition. Later, Anaxagoras of Clazomenae (c. 510 - c. 428 BC) a Greek philosopher of the 5th century BC wrote various texts on stellar topics. His work survived only in fragments quoted by later philosophers and writers. He attributed the cause of motion in the cosmos (gravity) to *Nous*, the mind. By later records of his work, the beginning is described as a mixture of ingredients, where the dense separated from the rare in a rotational motion leading to the formation of stars, the Sun, the Moon, etc. As per the quotation:

> "First it began the rotation from a small beginning, then more and more was included in the motion, and yet more will be included. Both the mixed and the separated and distinct, all things mind recognised. And whatever things were to be, and whatever things were, as many as are now, and whatever things shall be, all these mind arranged in order; and it arranged that rotation, according to which now rotate stars and sun and moon and air and aether, now that they are separated. Rotation itself caused the separation, and the dense is separated from the rare, the warm from the cold, the bright from the dark, the dry from the moist." —*Anaxagoras, translation in The first philosophers of Greece, by Fairbanks, Arthur, 1864-1944.*

In early times, astronomy was based on observations and the foundation of physics were more philosophical nevertheless a quantitative science. As the father of natural philosophy, the Greek philosopher Aristotle (384 - 322 BC) left various texts covering themes such as physics, metaphysics, poetry, rhetoric, linguistics, etc. Aristotle believed that all concepts and knowledge were based on perception. In his natural history treatise, Aristotle undertakes the classification of living species according to groups for plants, organisms, and subgroups such as blood and bloodless, and polymorphic attributes depending on how they move, walk, swim or fly. Aristotle arguments for the sphericity of Earth are as follows: only a sphere can account for the tendency of matter to fall towards a center, only a sphere can explain the apparent shadow of the earth on the moon during eclipses, only a sphere can explain the new constellations appearing in the sky as we travel from north to south pole.

Eudoxus (c. 408 - c. 355 BC) is known for his contributions to geometry in particular the theory of proportions and a kind of method of exhaustion, as applied to compute the area of say a circle inscribed in a polygon of n sides (as example). Moreover, the definition of proportion in "Euclid Elements" is attributed to Eudoxus. Hipparchus is referring to some works of Eudoxus, in his memoir on astronomy.

As per the baby Sun, Earth was placed at the center and the Sun revolving around, a model proposed as a way to explain the rising and setting of the fixed stars. As such, Aristarchus (c. 310 - c. 230 BC) came up with a method to measure the distance to the Moon with respect to Earth's diameter from the duration of the lunar passage on the horizon and the shadow of Earth on the Moon. His measurement of the distance to the Sun from the measurement of the angular separation of the Sun and the Moon when Moon was half-illuminated, was way under estimates. Nevertheless, an accurate measurement of the distance to the Moon was made by Hipparchus in the 2nd century BC, by the observation of the Moon from two adjacent cities, a method known as parallax.

Eratosthenes (c. 276 BC - c. 195/194 BC) made significant contributions to astronomy and geography. He is best known for his calculation of the circumference of the Earth (for a given tilt of Earth's axis). With the help of a Sundial, Eratosthenes measured the tilt of the Earth's axis as reported by Ptolemy. Eratosthenes also deduced the length of a year as $365\frac{1}{4}$ days, and suggested the introduction of a leap day every four years as a parameter for his calculation of the distance between Syene and Alexandria, a prerequisite for today calendar. While Earth curvature could be estimated from the escape velocity of a vessel on the horizon, Eratosthenes inferred the circumference of the Earth from a method based on the assumption the Sun was so far away that its rays were parallel and with a knowledge of the distance between two adjacent cities.

Hipparchus (c. 190 - c. 120 BC) was inspired by the precision of Babylonian astronomy. He probably compiled a list of astronomical observations from various sources such as ephemerides, lunar motion and eclipses, and using trigonometry developed some methods for the prediction of solar eclipses. He also calculated the distance from the Earth to the Moon from an eclipse of the Sun at the alignment of the Moon as viewed from Earth, a slightly different method than Aristarchus, leading to the Earth-Moon distance estimation of 60 Earth radii [6]. Hipparchus also contributed to the precession of the Earth's rotational axis.

Figure 11: Ptolemy's cartography (replica, c. 150 AD).

The Almagest by Claudius Ptolemy (c. AD 100 - c. 170) is one of the books on astronomy which has survived the flood. According to Ptolemy model, the universe is a sphere having for center Earth, a prerequisite for the establishment of the spherical coordinate system. To describe planetary motions, Ptolemy considered each planet on its own, moving in circles rotating in a larger circle (the deferent). By the epicycles of planetary motions, he achieved a description of the motion of planets in the solar system. Ptolemy contributed to various developments including applications to geography, such as maps in use during the Ptolemaic dynasty (see Fig. 11). As per Ptolemy's cosmography, the relative positioning of five planets of the solar system: Mercury, Venus, Mars, Jupiter and Saturn respectively, is depicted in some Latin scriptures of the early 16th century, at the time of Copernicus, e.g. Peter Apian, etc.

7 Astronomy in the Middle Ages and the Renaissance

The Middle Ages was a transition period in Western Europe between the end of the Roman era (at the fall of the 5th century AD) and the Renaissance (14th to 17th century AD). This period was marked by the ruling of the Church. The so-called geocentric model was viewed as a basement of the Church. Nevertheless, during the Renaissance the tendency was to challenge the cosmological views prevailing during that period. The Florentine influence of the meridional sphere was marked by scholars and painters of Christian heritage as Michelangelo, Giovanni Pico della Mirandola, etc. see the Heaven's vault, Sistine Chapel (Fig. 12).

Some of the great personalities from that period who had a preponderant influence on cosmological views are enumerated further down.

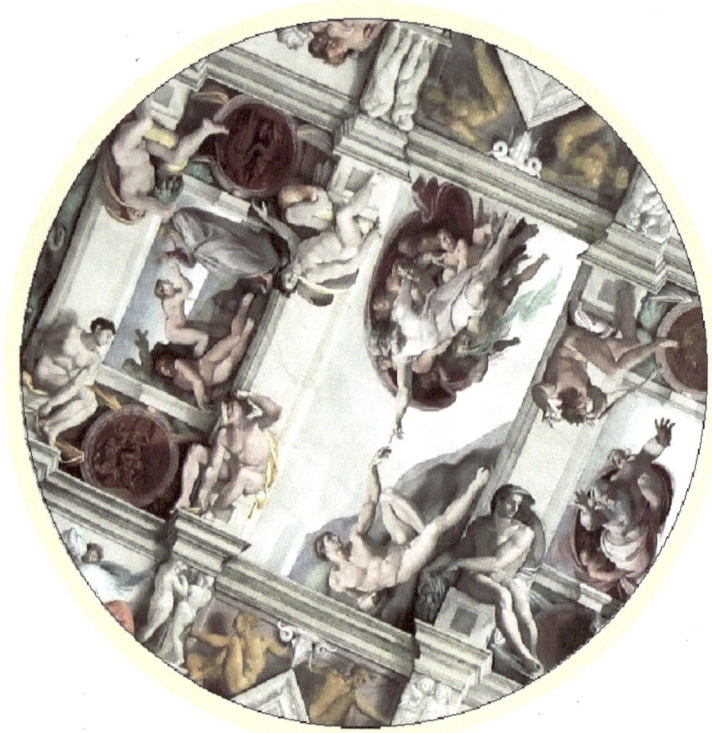

Figure 12: Michelangelo painting of Sistine Chapel ceiling (ca. 1508 - 1512) – *His faith and cosmological views, the holy worlds of great Angels.*

Nicolas Copernicus (1473 - 1543), placed the Sun rather than the Earth at the center. Copernicus was an educated man; he pursued his studies in Italy, before coming back to his country, present day Poland. In 1512, he moved to Frauenberg, where he was appointed to an administrative position in the Church. Copernicus wrote a reference book about his cosmological views on the revolution of the Heavenly Spheres. His legacy was published in 1543, no sooner before he died.

Giordano Bruno (1548 - 1600) was a Dominican friar and philosopher. He developed the idea of heliocentrism in a infinite universe. He proposed that the stars in the sky were distant Suns surrounded by planets, and that there would be a multitude of Earth like planets harbouring life. His view on cosmic pluralism and that there may be many inhabited worlds in the universe, was contrasting with the holy view of the Church. He wrote several books on bright and bold philosophical ideas, but his work was considered beyond his time. After eight years of trials on charges of heresy, the Church found him guilty and he was exhumed at the stake in Rome's Campo de Fiori, marking the end of the 17th century.

Tycho Brahe (1546 - 1601), a Danish nobleman, had an interest in astronomical observations. He is known for the Tychon system, which is a combination of the Copernican and Ptolemy models. According to his views, the Moon was orbiting around the Earth, and the planets around the Sun. For consistency with Ptolemy, he kept the Sun revolving around Earth. Due to a disagreement with the new Danish king, he moved to Prague in 1597. He had an observatory built in Benátky nad Jizerou, where he was assisted by Johannes Kepler. Fig. 13 is part of a mural quadrant, representing Tycho Brahe pointing at the sky, as viewed from a spyglass.

Galileo Galilei (1564 - 1642) played a significant role in observational astronomy. In parallel, Galileo became interested in the fall of bodies and gave birth to the first formulation of the principle of inertia. In May 1609, Galileo received a letter from Paolo Sarpi telling him about a spyglass that a certain Deutschman had constructed. The patent which was deposited in 1608 was attributed to the eye-glass maker Hans Lippershey. Soon after the news was released, Galileo began his endeavour to craft a telescope as inspired by the Dutchman's spyglass. As Galileo saw potential applications for his invention, he arranged a demonstration of the telescope to the Venetian Senate for its manufacturing. When Galileo directed his telescope to the night sky, he made amazing discoveries about the solar system. He could observe the phases of Venus giving progressive lighting appearances due to the rotation of the planets around the Sun. His work depicts the four largest satellites of Jupiter and he made astonishing observations about solar sunspots. He spoke with his editor. This was in 1623, when Galileo went to visit the new pope, Urban VIII, inquiring him about the writing of a book on heliocentrism. The pope granted him the authorization to write the book provided not to advocate heliocentrism, but to present a balanced view with the pros and cons of both heliocentrism and geocentrism. The book which presents the dialogue between Salviati who is a proponent of the Copernican system and Simplicio a proponent of geocentrism, was finalized in 1630 and published in 1632. Shortly after, as Galileo's writings were banned, he was convened to appear in the high court of Rome. The sentence has been alleviated allowing Galileo to pursue some of his work at the observatory.

Johannes Kepler (1571 - 1630) known for his planetary motion marked a turning point to the end of geocentrism. As Kepler noticed a tiny eccentricity of the Earth rotation around the Sun, he announced the below three axions known as the Kepler's laws : (i) The path of planets around the Sun is elliptical in shape, with the center of the Sun being located at one of the two foci, (ii) an imaginary line drawn from the center of the Sun to the center of the planet sweeps equal areas during equal intervals of time, (iii) the square of the orbital period of a planet is proportional to the cube of the semi-major axis of its orbit. While the first two laws are found in Astronomia Nova a book by Johanne Kepler in 1906, the third law was made public no later than in 1619. Whereas the seasons are attributed to the tilt of the Earth's axis, the third Kepler law was made in the spirit of Newton's law. The Kepler's laws are now part of the foundations of astronomy.

Figure 13: Element of mural quadrant engraving from Brahe's book: *Astronomiae Instauratae Mechanica (1598)*.

1 References

[1] BURNOUF, E.: *Mémoire Sur Deux Inscriptions Cunéiformes Trouvées Près D'Hamadan Et Qui Font Maintenant Partie Des Papiers Du Dr Schulz (French Edition)*. Nabu Press, 2010

[2] HESIOD: *Theogony*. The Perfect Library, 1914

[3] LAËRCE, D.: *Vies et Doctrines des Philosophes Illustres (French translation)*. Le Livre de poche, 1999

[4] MAGLI, G.: *Archaeoastronomy: Introduction to the Science of Stars and Stones*. Springer; 2nd ed., 2020

[5] OSSENDRIJVER, M.: *Babylonian Mathematical Astronomy: Procedure Texts*. Springer, 2012

[6] PANNEKOEK, A.: *A History of Astronomy*. Dover Publications Inc., 1990

[7] POOLE, R.M.: What became of the Taíno ? In: *Smithsonian Magazine* 42 (2011), S. 58–70

[8] REV A.H.M. A SATYE: *Astronomy and Astrology of the Babylonians.* Kessinger Publishing's, 2003

[9] SQUIER, E.G.: *Peru, Incidents of travel and exploration in the land of the Incas.* MacMillan and Co, London, 1877

Kelvin Universe - Part no. 2

Astronomy and Quantum Astrophysics

Centorus Publishing, LTD

The Etherington's reciprocity theorem and elementary derivation from wave theory

Yuri Heymann, Fall 2020.

Abstract The Etherington's reciprocity theorem, which is depicted in the relation between the luminosity distance of standard candles and the angular-diameter distance, known as the cosmic distance-duality, is a cornerstone in astrophysics for distance measurements at cosmic scales. This relation has been scrutinized by various research teams around the globe, as recent developments made it possible to probe this equation through observations such as X-ray surface brightness and the Sunyaev-Zel'dovich effect of galaxy clusters. The present work presents the views of the author and an elementary derivation of the cosmic distance duality in the context of wave theory and Hubble's parameter.

1 Introduction

The work of George F.R. Ellis on the *reciprocity theorem* shall be viewed as an abstract layer intended to be of general purpose and that can be implemented within the framework of general relativity. The core idea in [6] relates to the flux from a point source, which is observed along a bundle null geodesic with a small solid angle $d\Omega_s$ at the source and cross-sectional area dS_0 at the observer. Moreover, Ellis defines the area distance r_s^2 by the relation $dS_0 = r_s^2 d\Omega_s$. By the surface brightness, the flux is related to the solid angle of the source as follows:

$$F\, dS = \frac{L_0}{(1+z)^2} \frac{d\Omega}{4\pi}\,, \qquad (1)$$

where F is the flux, L_0 is a function of the intrinsic luminosity of the source L_s, and z is the redshift. It follows that the luminosity distance D_L is tied to r_s by the relation $D_L = (1+z)\, r_s$, as this is the connection between (1) and the flux expressed as $F = \frac{L_0}{4\pi D_L^2}$. The reciprocity theorem involves the dual relation $dS_s = r_0^2 d\Omega_0$ resulting from the transposition of the roles of the source and the observer, leading to the quadratic term $(1+z)^2$ in the cosmic duality equation:

$$D_L = (1+z)^2 D_A\,, \qquad (2)$$

where D_L is the luminosity distance and D_A is the angular-diameter distance. This is the well-known distance-duality equation referred to in [8]. In his work, Etherington investigated the relations amongst cosmological distances, while referring to his communication with Tolman regarding model independent testing of the cosmic distance duality.

The view presented in [2], is that the reciprocity theorem remains valid when photon number is conserved and gravity is described by a metric theory with photons travelling on null geodesics. In their study, different estimates of angular-diameter distance for a set of redshifts coming from sources such as radio galaxies, compact radio sources and X-ray clusters were used as standard rulers conjointly with type Ia supernovae for the luminosity distance, as means to probe the cosmic distance duality. While the standard approach to probe the luminosity distance is from standard candles, a way to determine the angular-diameter distance is from the observation of clusters of galaxies through X-ray surface brightness and the Sunyaev-Zel'dovich effect in [20]. A valuable source regarding the latter is depicted in [21], showing no significant violation of the cosmic distance duality. Interestingly, in this study, the clusters deviating most from the expected behavior were those showing bimodal structures in the data. In the work of [3], the ratio between the luminosity distance and the angular-diameter distance times the quadratic term for the surface brightness, which is defined as $\eta(z) = \frac{D_L}{D_A(1+z)^2}$, was bound to be 1.01 ± 0.07 at 68% confidence level. Notwithstanding, the reliability of confidence bands reported in many studies is often questionable due to model misspecifications or systematic errors not accounted for in the variance of the estimates. Other aspects, such as performing analyses with sparse datasets, are highlighted in [15] as part of the cosmic duality research involving topics in particle physics and opacity, e.g.[1]. Extensive calibration work aiming at ensuring the quality of the measurements involved in the cosmic duality had to be performed. As such, sensitivity analyses across different samples of galaxy clusters were done in [9]. The deviation of the distance duality in terms of the cosmic absorption is the main issue in [13], more specifically for supernova dimming.

Although the school of thought was that violation of the distance duality would bring new physics, among exogenous effects, it was suggested that extinction by a diffuse effect through the so-called cosmic dust of the inter-galactic medium could affect the determination of the luminosity distance [5]. Moreover, modeling aspects impacting the determination of the angular-diameter distance would substantially alter the outcome of the distance-duality relation. As such, improper modeling of the 3-D gas density profile in galaxy clusters may induce statistical errors in proportion exceeding acceptance level, producing significant departures from the cosmic distance duality as known today, see [14]. Nonetheless, all of the methods developed to probe cosmic distance measurements rely to some extent, on some sort of cosmological ladder. The idea was to use the distance determined for nearby objects with well-specified methods as a benchmark to calibrate other methods for the distance of farther objects. Though involving a significant amount of modeling, the use of gravitational lensing as a standard ruler to probe the cosmic distance duality is among the topics in this field as shown in [10, 12, 17]. Consistency across methods and observations is key for metrological aspects related to the cosmological ladder, whether to enhance existing or future methods. While the estimation of Hubble's parameter was the main benchmark for quantitative methods in cosmology, the current trend is to calibrate models on distance measurements directly as per the cosmic duality. The present work offers a simple method for the construction of calibration curves for distance measurements, which can be used as a collaborative tool to compare modeling approaches.

The earlier work of the author on cosmic distance measurements is available in Progress in Physics, a quarterly bulletin founded by Dmitri Rabounski, Florentin

Smarandache, and Larissa Borissova in 2005. In the below, the derivation of the cosmic distance duality in the context of wave theory and Hubble's constant as a single parameter is made available by the author. It appears there is an implicit connection between the present work and the reciprocity theorem by Ellis, G.F.R., as "the core of the reciprocity theorem is the fact that many geometric properties are invariant when the roles of the source and observer in astronomical observations are transposed" [7]. This annotation highlights interesting aspects about the cosmic distance duality as shown further down when modeling the light wavefront with respect to the source and the observer. Other research lines on the cosmic duality include the recent work of [18], which modeling approach is based on the electrical paradigm as per Gauss's and Poynting's theorems. The current derivation of the distance measurements as a prerequisite for the cosmic duality is detailed below.

2 Galaxy clusters as standard rulers and some of the principles in X-ray observational astronomy

X-ray are electromagnetic radiations at short wavelengths ranging from a picometer to about ten nanometers, of interest in observational astronomy such as for the study of the corona of stars and nebulae, X-ray jets emitted from black holes, or the inter-galactic medium. To escape from the opacity skew of the atmosphere, X-ray telescopes had to be placed in space. Early X-ray surveys included Skylab in the 70s for imaging of the solar corona. Significant breakthroughs in observational astronomy were possible with Chandra X-ray observatory launched by the NASA in 1999, in parallel to XMM-Newton by the ESA. Later, came NuStar launched in 2012 for a survey of compact objects and black holes, and many more others. With prominent developments in optical geometry and modeling aspects related to metrology, electrodynamics and gravitation, new methods to probe the size of galaxy clusters were made possible through X-ray observations.

A brief overview of the method described in [21], to probe the cosmic distance duality through the latest astronomical observations is described here. The luminosity distance is obtained from type Ia supernovae conjointly with the distance modulus for a set of redshifts. The challenge is to measure the angular-diameter distance as the actual size of astronomical objects is not known precisely. This is done for galaxy clusters using measurements from X-ray surface brightness and the Sunyaev-Zel'dovich effect (SZ). Galaxy clusters contain large quantities of hot and ionized clouds of gas, emanating from the corona of stars and nebulae at high temperatures between 10 to 100 megakelvins. This hot gas radiates in the X-ray domain through bremsstrahlung, which are radiations produced by the deceleration of charged particles when deflected by other particles. This intra-cluster gas distorts the Cosmic Microwave Background Radiations (CMBR), through the SZ effect. An explanation that was proposed for the SZ effect, is the inverse Compton interaction of photons receiving an energy boost when colliding with high energy free electrons. The SZ effect is referring to a skew in the CMBR spectrum, corresponding to a dimming of the brightness at low frequencies but increased at high frequencies. The drop in temperature or brightness of the CMBR spectrum in the Rayleigh-Jeans region due to the SZ effect, is a function of electron temperature and density. The X-ray surface brightness is expressed as a function of the volume of the cluster and electron temperature and density. Thus, one can eliminate the electron density term and estimate the core radius of the clusters, as outlined in [19].

The effect of a set of explanatory variables for the SZ effect and X-ray surface brightness have been considered, such as the shape and distribution of the gas cloud forming the clusters, non-isothermality, and adjustments such as Doppler effect fom the cluster peculiar velocity, etc. The isothermal β-model as described in [11] is a parametric model involving a density profile of the cluster, expressed as a function of the ratio of a variable r representing a parametric radius ranging from zero to infinity divided by the core radius of the cluster r_c, the unknown of the problem, and an additional shape factor β. The fractional temperature decrement $\Delta T_{SZ}/T$ of the CMBR due to the SZ effect and the X-ray surface brightness S_x of the cluster of galaxies, are the two observables incorporated into a single equation as a basis for the model. The theoretical relations for both of these observables used conjointly with the density profile of the cluster, lead to an expression relating the core radius of the cluster to the above observables and the electron temperature. The electron temperature T_e is obtained by fitting the observed X-ray spectrum to the theoretical X-ray spectrum expected for an isothermal gas. Other aspects of the method including model calibration are discussed in [4], to probe the radius of the cluster. Lastly, the angular-diameter distance is obtained from the size of the cluster and angular size.

3 Mathematical foundation for cosmic distance measurements

3.1 Derivation from wave theory on the basis of wave-particle duality

Given a wave function $E(r,t) = A\cos(w\,t)$ moving in a radial direction, where A is the amplitude and w the wave frequency, the energy flux denoted S is equal to the time derivative of the square of the wave function. The square of the wave function is introduced to account for the three-dimensionality of space as a propagation medium for the wave. Hence, we get:

$$S = \frac{\partial}{\partial t} E^2(r,t) = w\,A^2\,2\,\sin(wt)\cos(wt), \tag{3}$$

which is the underlying equation for the wave-particle duality. The frequency is expressed as $w = \frac{c}{\lambda}$ where c is the speed of light and λ the wavelength. The effective value by the root-mean square, leads to the relation:

$$S = w\,A^2. \tag{4}$$

As the flux S is expressed in joules per second and square meters and the wave frequency w is an inverse of time, the quadratic term of the wave amplitude A^2 which is expressed in joules per square meters, represents a surface energy density. This equation can be related to the flux of an electric field as defined in Poynting's theorem by $S = c\,\varepsilon_0\,E^2$, where E is the standard wave of the electric field, and ε_0 the vacuum permittivity. Interestingly, the energy flux as shown in (4), can be decomposed into a product of the wave frequency and amplitude of the wave function.

In the remaining portion of the manuscript, the letter E is a variable representing the energy of photons as per Planck's formula $E = \frac{hc}{\lambda}$, where h is the Planck's constant, c the speed of light, and λ the wave frequency. This variable represents the frequency component of the energy flux from the quadratic wave function defined above. The change in the wave frequency of a light source causes the energy of photons to vary over time, which in the tired-light paradigm is explained by the effect of an extinction coefficient resulting from diffusion of dust in the inter-galactic medium. As we suppose that the number of cycles of the light wave is conserved, an apparent expansion of the luminous phase is produced as the result of photon energy dispersion. This expansion of the luminous phase, may be translated into an increase in the velocity of the light wavefront as photons travel through space (Fig. 1). For consistency with special relativity, a time contraction is introduced to the model to maintain the speed of light invariable.

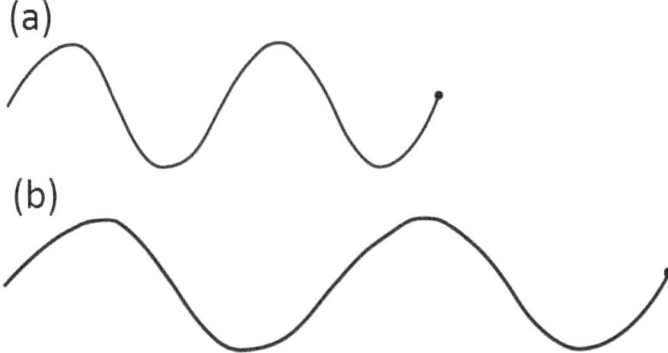

Figure 1: Where (a) represents a light wave, which is stretched as a fabric of space in (b). We can see that in (b), the light wavefront is moving faster than in (a).

The change in energies of photons between the time of emission at the source and the time when the observer is reached, leads to the below relation providing the definition of redshifts:

$$1 + z = \frac{E_s}{E_0}, \qquad (5)$$

where E_s is the energy of photons at the source, E_0 is the energy of photons when reaching the observer, and z is the redshift.

A first-order rate extinction is considered for the variation of the energy of photons travelling throughout space, which is expressed as follows:

$$\frac{\dot{E}}{E} = -H_0, \qquad (6)$$

where H_0 is the Hubble's constant expressed as a coefficient of extinction. This formulation contrasts with the standard definition of Hubble's constant, as a unit of measurement for the expansion of the universe. As per the first-order rate extinction in (6), the energy of photons as a function of time, follows an exponential law, which is expressed as follows:

$$E_t = E_0 \exp(-H_0 t), \qquad (7)$$

where E_t denotes the energy of photons at any instant of time t. Time t is defined as an automorphic variable which is elapsing over itself. Furthermore, time t is set equal to zero when photons reach the observer indicating the present. As such, time t is a negative number when referring to photons' energies in the past, and a positive number for photons' energies at any time interval in the future.

The reciprocal of (7) leads to the below relation between photons' energies at the source and at the observer, which expression is as follows:

$$E_s = E_0 \exp(H_0 T), \qquad (8)$$

where T denotes the light-travel time separating the time zone of the source from the observer. The letter E_s stands for the energy of photons at the source, whereas E_0 for photons' energy when they reach the observer.

The below analyses are performed in two steps. First, a time-varying light wavefront to accomodate the apparent expansion of the luminous phase. Second, a time contraction is introduced to the model to keep the light wavefront at the speed of light.

3.1.1 Light wavefront with respect to the source

The light-wavefront velocity as viewed from the source before the application of time contraction, is expressed as follows:

$$v(t) = c \frac{E_s}{E_t}, \qquad (9)$$

where E_s is the energy of photons when emitted at the source, and E_t is the energy of photons as a function of the automorphic variable of time t. To keep the light wavefront at the speed of light, the below time contraction is applied to the wave function:

$$\frac{\delta t_s}{\delta t} = \frac{E_s}{E_t}. \qquad (10)$$

The light-travel time relative to the source, is expressed as follows:

$$T_s = \int_{-T}^{0} \frac{\delta t_s}{\delta t} dt = \int_{-T}^{0} \frac{E_s}{E_t} dt. \qquad (11)$$

The above integral follows from the definition of time as an automorphic variable, allowing the application of the partial derivative to the time variable over itself, as an operator in the chain rule.

By substitution of (7) into (11) and integrating, we get:

$$T_s = \frac{E_s}{E_0} \frac{1}{H_0} \left(1 - \exp(-H_0 T)\right). \qquad (12)$$

By substitution of (8) into (12), we get:

$$T_s = \frac{E_s}{E_0} \frac{1}{H_0} \left(1 - \frac{E_0}{E_s}\right). \qquad (13)$$

By substitution of (5) into (13), we obtain:

$$T_s = \frac{z}{H_0}, \qquad (14)$$

which is the light-travel time measurement for the luminosity distance, where z is the redshift and H_0 the Hubble's constant.

3.1.2 Light wavefront with respect to the observer

The light-wavefront velocity as viewed from the observer before application of the time contraction, is expressed as follows:

$$v(t) = c \frac{E_0}{E_t}. \qquad (15)$$

where E_s stands for the energy of photons at the time of emission at the source, and E_t is the energy of photons at any instant of time, as defined by the variable t. To keep the light wavefront at the speed of light, the below time contraction is applied to the wave function:

$$\frac{\delta t_0}{\delta t} = \frac{E_0}{E_t}. \qquad (16)$$

The light-travel time relative to the observer, is expressed as follows:

$$T_0 = \int_{-T}^{0} \frac{\delta t_0}{\delta t} dt = \int_{-T}^{0} \frac{E_0}{E_t} dt. \qquad (17)$$

The above integral follows from the definition of time as an automorphic variable as seen in (11), by assuming a monochromatic wave.

By substitution of (7) into (17) and integrating, we get:

$$T_0 = \frac{1}{H_0} \left(1 - \exp(-H_0 T)\right). \qquad (18)$$

B substitution of (8) into (18), we get:

$$T_0 = \frac{1}{H_0} \left(1 - \frac{E_0}{E_s}\right). \tag{19}$$

By substitution of (5) into (19), we obtain:

$$T_0 = \frac{1}{H_0} \frac{z}{(1+z)}, \tag{20}$$

which is the light-travel time measurement for the proper distance, where z is the redshift and H_0 Hubble's constant.

3.2 Derivation from homogeneous expansion in the 3-D representation of the luminous phase

The luminous expansion is a paradigm, in which Hubble's constant represents a homogeneous rate of expansion of the luminous phase as a fabric of space. In the below, we provide the derivation of the distances measurements in this context specifically.

3.2.1 Luminosity distance

The luminosity distance refers to the theoretical distance as obtained from the brightness of standard candles with the inverse-square law. As such, type Ia supernovae are considered standard candles, meaning they all have the same absolute brightness when they explode. From their apparent brightness, we can deduce the luminosity distance, as their apparent brightness decreases by the inverse-square law as follows:

$$F_{obs} = \frac{L_0}{4\pi r_L^2}, \tag{21}$$

where F_{obs} is an observable. In (21) above, L_0 is the luminosity of the source, and r_L the luminosity distance. As a variable, L_0 is connected to the intrinsic luminosity L_s by a function, which may account for adjustments such as the change in wavelength of photons when photon count is conserved. The calculation of the luminosity distance involves a formula for the distance modulus defined as the difference between the apparent and absolute magnitude of an astronomical object. The magnitude system is a logarithmic scale of the brightness as defined in [16], where the absolute magnitude represents the magnitude of the source at a theoretical distance of ten parsecs.

By considering a photon emitted from a supernova, moving away from the center in a radial direction, the luminosity distance is described by the below differential equation:

$$\frac{dr_L}{dt} = c + H_0 r_L, \tag{22}$$

where r_L is the luminosity distance, H_0 the Hubble's constant, and c the speed of light. This equation expresses the rate of increase in the luminosity distance as a function equal to the speed of light plus an expansion term corresponding to the luminosity distance times the Hubble's constant.

The solution to this differential equation is obtained by combining both the homogeneous and particular solutions of the above, with initial conditions $r_L = 0$ when $t = 0$ and terminal time T, leading to:

$$r_L = \frac{c}{H_0}\left(\exp(H_0 T) - 1\right). \tag{23}$$

By definition, the rate of increase of the scale factor a is given by $\frac{da}{dt} = H_0\, a$, leading to $dt = \frac{da}{H_0 a}$. Furthermore, the relation between the scale factor a and the redshift z is given by the cosmological redshift equation $(1+z) = \frac{a_0}{a_t}$, where the scale factor at present time a_0 is equal to one. Thus, the light-travel time versus redshift is expressed as follows:

$$T = \int_{1/(1+z)}^{1} \frac{da}{H_0 a} = \frac{1}{H_0}\ln(1+z). \tag{24}$$

Eqs. (23) and (24) yield:

$$r_L = \frac{c}{H_0} z, \tag{25}$$

which is the expression of the luminosity distance r_L as a function of redshift z, where c is the speed of light and H_0 the Hubble's constant. This equation is connected to (14) by the relation $r_L = c T_s$, as a yardstick expressing a unit of time into metric scales.

3.2.2 Euclidean and proper distance

A measurement of the distance is obtained by calculating the corresponding distance as if universe was not expanding, which is the proper distance viewed in Euclidian 3-D geometry. Let us introduce the variable y to this distance measurement. By considering a photon moving towards the observer, we have:

$$\frac{dy}{dt} = -c + H_0 y, \tag{26}$$

which is the differential equation describing the time evolution of the proper distance separating a photon from the observer. This equation expresses the rate of decrease in the proper distance, as a function equal to the expansion term expressed as the proper distance times Hubble's scale, from which the speed of light is subtracted. By setting time zero at a reference T_b in the past, we have: $t = T_b - T$, where T is the light-travel time separating the photons from the observer. Thus, we have $dt = -dT$, leading to:

$$\frac{dy}{dT} = c - H_0 y, \tag{27}$$

where the boundary condition is given by $y(T = 0) = 0$ when photons reach the observer. The solution to this differential equation is obtained by combining both the homogeneous and particular solutions of the above, with time between 0 and T, leading to:

$$y = \frac{c}{H_0}\left(1 - \exp(-H_0 T)\right), \qquad (28)$$

where y represents the proper distance separating the observer from the source, and T, the light-travel time separating the source from the observer. By substitution of (24) into (28), we obtain:

$$y = \frac{c}{H_0}\frac{z}{(1+z)}, \qquad (29)$$

which is the expression of the proper distance as a function of redshift z, where c is speed of light and H_0 the Hubble's constant. This equation is connected to (20) by the speed of light, i.e. $y = c T_0$.

3.3 The cosmic distance duality

From (25) and (29), we have:

$$r_L = (1+z)\, y, \qquad (30)$$

where r_L is the luminosity distance, y the proper distance, and z the redshift.

The angular-diameter distance d_A of an object, which is defined in terms of the object's actual size x, and the angular size θ of the object as viewed from Earth, is expressed as follows:

$$d_A = \frac{x}{\theta}. \qquad (31)$$

From the expansion of the luminous phase as viewed as a fabric, the apparent size of celestial objects is stretched by a factor (1+z), and the apparent angular size increased by the same factor, leading to the below relation between the actual distance y and the angular-diameter distance d_A:

$$y = (1+z)\, d_A. \qquad (32)$$

Eqs. (30) and (32) yield:

$$r_L = (1+z)^2 d_A, \qquad (33)$$

where r_L is the luminosity distance, d_A the angular-diameter distance, and z the redshift. This is the well-known cosmic distance-duality equation, which we have derived above in the context of wave theory, on the basis of wave-particle duality and as a particular expansion in a fabric of space.

4 Conclusion

An elementary derivation of the cosmic distance duality in the context of wave theory and Hubble's constant, is disclosed in the current manuscript. Although, the approach undertaken by the author in this context deviates from heuristics commonly in use in cosmology, it is quite critical and has the advantage of revealing the cosmic duality in its most simplistic form. The connection between the present work and the reciprocity theorem by Ellis, G.F.R. is established throughout the development presented in this study. The reciprocity theorem depicted in the relation between the angular-diameter distance and luminosity distance of standard candles, known as the cosmic distance duality, is well documented in the literature and is one of the valuable tools in observational astronomy for its use among other things, such as the development of measurement devices applied to metrology, cartography, and surveying. All of the methods developed to probe cosmic distance measurements rely to some extent, on the so-called cosmological distance ladder. The rationale is to use the distance determined for nearby objects with appropriate methods as a benchmark to calibrate other methods for the distance of farther objects. The key to the cosmological ladder is consistency across methods and observations, and a fair amount of detailed work.

5 References

[1] AVGOUSTIDIS, A. ; BURRAGE, C. ; REDONDO, J. ; VERDE, L. ; JIMENEZ, R.: Constraints on cosmic opacity and beyond the standard model physics from cosmological distance measurements. In: *Journal of Cosmology and Astroparticle Physics* 10 (2010), S. 024

[2] BASSETT, B. A. ; KUNZ, M.: Cosmic distance-duality as a probe of exotic physics and acceleration. In: *Physical Review D* 69 (2004), S. 101305

[3] BERNARDIS, F. ; GIUSARMA, E. ; MELCHIORRI, A.: Constraints on dark energy and distance duality from Sunyaev-Zel'dovich effect and Chandra X-ray measurements. In: *International Journal of Modern Physics D* 15 (2006), Nr. 5, S. 759–766

[4] BIRKINSHAW, M. ; HUGHES, J. P. ; ARNAUD, K. A.: A measurement of the value of the Hubble constant from the X-ray properties and the Sunyaev-Zel'dovich effect of Abell 665. In: *The Astrophysical Journal, Part 1* 379 (1991), S. 466–481

[5] CORASANITI, P. S.: The impact of cosmic dust on supernova cosmology. In: *Monthly Notices of the Royal Astronomical Society* 372 (2006), S. 191–198

[6] ELLIS, G. F. R.: *Relativistic cosmology*. Proceedings of the International School of Physics "Enrico Fermi", Course 47: General relativity and cosmology, edited by R. K. Sachs. Academic Press, New York and London, pp. 104-182, 1971

[7] ELLIS, G. F. R.: On the definition of distance in general relativity: I.M.H. Etherington (Philosophical Magazine ser. 7, vol. 15, 761 (1933)). In: *Gen. Relativ. Gravit.* 39 (2007), S. 1047–1052

[8] ETHERINGTON, I. M. H.: On the definition of distance in general relativity. In: *Philosophical Magazine series 7* 15 (1933), S. 761–773

[9] GONÇALVES, R. S. ; HOLANDA, R. F. L. ; ALCANIZ, J. S.: Testing the cosmic distance duality with X-ray gas mass fraction and supernovae data. In: *Monthly Notice Letters of the Royal Astronomical Society* 420 (2012), S. L43–L47

[10] HOLANDA, R. F. L. ; BUSTI, V. C. ; ALCANIZ, J. S.: Probing the cosmic distance duality with strong gravitational lensing and supernovae Ia data. In: *Journal of Cosmology and Astroparticle Physics* 2016 (2016), S. 54

[11] INAGAKI, Y. ; SUGINOHARA, T. ; SUTO, Y.: Reliability of the Hubble-Constant Measurement Based on the Sunyaev-Zel'dovich Effect. In: *Publications of the Astronomical Society of Japan* 47 (1995), S. 411–423

[12] LIAO, K. ; LI, Z. ; CAO, S. ; BIESIADA, M. ; ZHENG, X. ; ZHU, Z. H.: The Distance Duality Relation from Strong Gravitational Lensing. In: *The Astrophysical Journal* 822 (2016), S. 74

[13] LIMA, J. A. S. ; CUNHA, J. V. ; ZANCHIN, V. T.: Deformed Distance Duality Relations and Supernova Dimming. In: *The Astrophysical Journal Letters* 742 (2011), Nr. 2, S. L26

[14] MENG, X. L. ; ZHANG, T. J. ; ZHANG, H. ; WANG, X.: Morphology of Galaxy Clusters: A Cosmological Model-independent Test of the Cosmic Distance-Duality Relation. In: *The Astrophysical Journal* 745 (2012), S. 98

[15] NAIR, R. ; JHINGAN, S. ; JAIN, D.: Cosmic Distance Duality and Cosmic Transparency. In: *arXiv: astro-ph/1210.2642* (2012)

[16] POGSON, N.: Magnitudes of Thirty-six of the Minor Planets for the first day of each month of the year 1857. In: *Monthly Notices of the Royal Astronomical Society* 17 (1857), S. 12–15

[17] RUAN, C. Z. ; MELIA, F. ; ZHANG, T. J.: Model-independent Test of the Cosmic Distance Duality Relation. In: *Astrophysical Journal* 866 (2018), S. 31

[18] SCHULLER, F. P. ; WERNER, M. C.: Etherington's Distance Duality with Birefringence. In: *arXiv: gr-gc/1707.0126* (2017)

[19] SILK, J. ; WHITE, S. D. M.: The Determination of q_0 Using X-Ray and Microwave Measurements of Galaxy Clusters. In: *The Astrophysical Journal Letters* 226 (1978), S. L103

[20] SUNYAEV, R. A. ; ZEL'DOVICH, Ya. B.: The Observation of Relic Radiation as a Test of the Nature of X-Ray Radiation from the Clusters of Galaxies. In: *Comm. Astrophys. Space Phys.* 4 (1972), S. 173–178

[21] UZAN, J. P. ; AGHANIM, N. ; MELLIER, Y.: Distance Duality Relation from X-Ray and Sunyaev-Zel'dovich Observations of Clusters. In: *Physical Review D* 70 (2004), S. 083533

Probing cosmic transparency from the zCosmos deep-field galaxy survey

Yuri Heymann, winter 2020.

Some of the content in the present text are borrowed from Heymann, Y. (2014), Progress in Physics, Volume 10, Issue 4, pp. 217-221.

Overview The scope of the present text is the probing of cosmic transparency as viewed from a telescope imaging, giving rise to some galaxy-redshift distributions. Such survey typically consists of a collection of galaxies sorted by their respective redshifts, as observed for a given spectroscopic area of the celestial sphere. As Hubble law unveils some relation between redshifts and distances, a curve representing a distribution of redshifts versus distances is implied from the galaxy survey. The slicing of a vision of the bundle of an aperture from a telescope imaging into tiny redshift buckets, leads to some galaxy density profile (i.e. some curves expressing galaxy densities versus light-travel time) as a result of a measurement of galaxy distancing. Various effects having an impact on galaxy-redshift distributions are under consideration such as obscuration by foreground galaxies, geometrical skews such as geodesics followed by light rays in curved spacetime, optical resolution in far infrared, etc. As a benchmark, some theoretical galaxy density profiles have been generated by simulation techniques. Notwithstanding, theoretical approximations of galaxy-redshift distributions are worth considering. The best fitting of survey galaxy distribution against benchmark, is one of the approaches providing some insights about cosmic transparency and expansion scenarios.

1 Introduction

The spreading of computer simulations is a promising and powerful ingredient with various applications in astrophysics. As an example, the Millennium Simulation project carried out at the Max Planck Institute for Astrophysics, is amongst the largest n-body simulation carried out so far for simulations of the formation of large structures in the universe with the help of a cluster of 512 processors.

> Our rationale was to slice a galactic survey into small redshift buckets. We then used cosmological models to compute the volume of each bucket and derived the galactic density curve versus the redshift, or light-travel time. We used the simulation to generate a uniform distribution of galaxies for each redshift bucket. We then computed the number of visible galaxies (i.e., those that were not covered by foreground galaxies) to derive a simulated galactic density curve. Our method requires only a cosmological model, a behavior for the galactic density, and the average galactic radius versus the redshift. —*Assistant laborantin Constantine Youssef, 2014.*

The below is based on Hubble law, as a principle underpinning the computation of cosmic distances from redshifts. Moreover, the perfect cosmological principle stating that the universe is isotropic on a large scale is the underlying axiom in use to materialize the galaxy-density versus redshift distribution.

The model employed in the present work for the computation of proper distance separating Earth from distant light sources, is borrowed from above text respective to the cosmic-distance duality derived in context of wave theory as per Ellis's reciprocity theorem. As such, the proper distance between the observer and a source is expressed as follows:

$$r = \frac{1}{H_0} \frac{z}{(1+z)}, \qquad (1)$$

where r is the proper distance expressed as a light-travel time separating Earth from the source (say a Galaxy), H_0 Hubble's constant and z the redshift.

Various aspects of the galaxy-redshift distribution are covered in the remaining portion of the manuscript such as sampling, simulation techniques, and modeling aspects including obscuration by the some galactic dust, common assumptions, scenario analysis, etc. Lastly, the expected galaxy density profiles (as a density versus time scale expressed as light-travel time), is displayed in Fig. 3 (Third section) showing the density curves in three distinct scenarios, the Squash, Go-Bang, and Luminous Universe respectively. Moreover, the relation between cosmology and today particle physics, is drawn in the conclusion.

2 Methodological framework

2.1 Galaxy sampling in deep field

Say the zCosmos deep-field galaxy survey consists of a collection of visible galaxies with respective redshifts obtained for a given spectroscopic area in the sky spanning wavelengths from the visible spectrum to the deep infrared [1]. By slicing the survey into tiny redshift buckets and counting galaxies, the galaxy density is determined by a ratio of galaxy counts to the volume of a bucket. The volume of a bucket between the slice of the celestial sphere contained within the lower and upper radii of the bucket is scaled by the ratio of the survey spectroscopic area to the solid angle of the sphere.

As viewed from Earth, the volume of a slice of the celestial sphere is simply: $V_i = \frac{4\pi}{3}\left(r_i^3 - r_{i-1}^3\right)$, where r_{i-1} and r_i are the lower and upper radii of the bucket.

The spectroscopic area of the zCosmos galaxy survey ϕ_{surv} was determined to be 0.075 square degrees as per chart (see Fig. 1), as a measure of the solid angle of the sky opening viewed from a telescope determined from the area of the projection by the right ascension in degrees and $180/\pi$ times the sine of the declination. The above leads to the below ratio of the survey spectroscopic area to the solid angle of the sphere, expressed as follows:

$$\eta_{surv} = \frac{\phi_{surv}}{4\pi(180/\pi)^2}. \qquad (2)$$

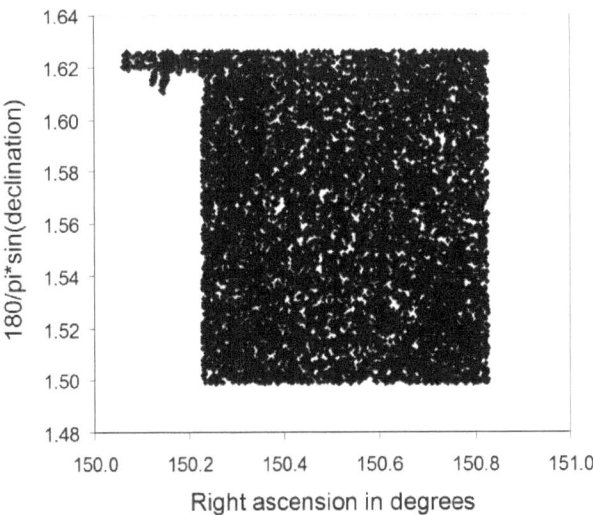

Figure 1: zCosmos galaxy survey deep field, carried out with the V.L.T. at the ESO Paranal Observatory under Programme ID: LP175.A-0839.

The volume of the ith bucket of the survey is equal to $\eta_{surv} V_i$, as a prerequisite for the computation of galaxy densities from counts and redshifts.

2.2 The simulation framework

Some of the requirements for the simulation of galaxy density profiles include the relation between local and observed galaxy radius and densities, resulting in expressions for galaxy-count densities and radius versus redshifts, which quantities are expressed in an aggregate basis. A standardized uniform redshift slicing (i.e. $\Delta z = 0.1$) was employed to compute the galaxy-density curve of the survey, say $z \in \{0, z_1, z_2, ..., z_n\}$, where $z_{i+1} = z_i + \Delta z$. For each iteration, say redshifts z_1 to z_n, N_i galaxies were generated from uniformly distributed galaxies as per the perfect cosmological principle (i.e. isotropic universe on a large scale). The position of each galaxy determined in the astronomical spherical coordinate system is given by (r, θ, φ), where r is the radial distance, $\theta \in [-\frac{\pi}{2}, \frac{\pi}{2}]$ is the declination, and $\varphi \in [0, 2\pi]$ is the right ascension.

Each galaxy has a radius which is used part of a criterion to determine whether such galaxy is visible amongst foreground galaxies. This criterion is based on some calculation of the tangential component of the distance separating nearby galaxies as viewed from a telescope imaging.

The spectroscopic area of the simulation itself is determined by the surface spanned by the declination and right ascension, say $\varphi \in [\varphi_{min}, \varphi_{max}]$ and $\theta \in [\theta_{min}, \theta_{max}]$. Say ϕ_{sim} is the spectroscopic area of simulation expressed as:

$$\phi_{sim} = \left(\frac{180}{\pi}\right)^2 (\sin\theta_{max} - \sin\theta_{min}) \times (\varphi_{max} - \varphi_{min}), \tag{3}$$

leading to a ratio of the spectroscopic area to the solid angle of the sphere, expressed as follows:

$$\eta_{sim} = \frac{\phi_{sim}}{4\pi(180/\pi)^2}. \tag{4}$$

The number of galaxies to be generated for a redshift bucket $[z_{i-1}, z_i]$, is computed as the volume V_i of a slice of the celestial sphere scaled by the spectroscopic area, to be multiplied by the galaxy density ρ_i, leading to:

$$N_i = \rho_i \, \eta_{sim} V_i, \tag{5}$$

where ρ_i the local galaxy density at redshift z_i, and η_{sim} and V_i as defined above (N_i being the number of galaxies to be generated for given redshift bucket i).

The position of such a galaxy is generated from two independent uniform random variables, say X and Y, as drawn from the equidistribution on a unit interval $[0, 1]$, leading to the declination and right ascension of galaxies, expressed as follows:

$$\begin{aligned}\theta &= \theta_{min} + X(\theta_{max} - \theta_{min}),\\ \varphi &= \varphi_{min} + Y(\varphi_{max} - \varphi_{min}).\end{aligned} \tag{6}$$

The new galaxies are attributed a radial distance, from the light-travel time at redshift z_i.

2.3 Criterion to determine whether a galaxy is visible

For each new galaxy, the below criterion is in use to determine whether it is visible or hidden by foreground galaxies. As an example, let us consider galaxy A and galaxy B, where galaxy A lies in the foreground (see Fig. 2).

The rationale is to compute the distance between the projection of galaxy A onto the plane of galaxy B, and galaxy B itself. This distance is referred further down as the tangential or projected distance between A and B (onto the plane of B). Say that such projected distance is larger than or equal to critical distance D_{crit}, then galaxy B is said to be visible; otherwise we say that galaxy B is hidden. The so-called projected distance between A and B, denoted $\mathrm{dist}(A, B)$ is computed as follows:

$$\mathrm{dist}(A, B) = \sqrt{(x_A - x_B)^2 + (y_A - y_B)^2 + (z_A - z_B)^2}, \tag{7}$$

where $\{x, y, z\}$ are the Cartesian coordinates of both galaxies A and B, projected onto the plane of B. Subscripts A and B are used as specifiers for the coordinates of galaxies A and B.

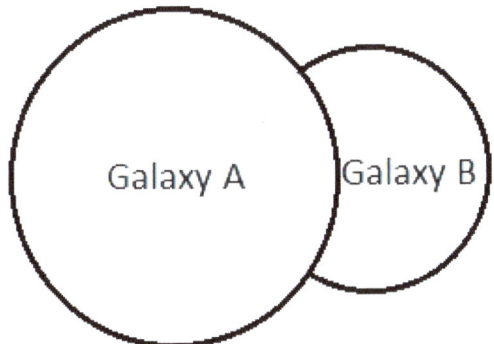

Figure 2: A foreground galaxy partially covering a more distant galaxy.

By the spherical coordinates, the positioning of a pair of galaxies A and B, expressed in Cartesian coordinates, is as follows:

$$\begin{aligned} x &= r_B \cos\theta \sin\varphi\,, \\ y &= r_B \cos\theta \cos\varphi\,, \\ z &= r_B \sin\theta\,, \end{aligned} \qquad (8)$$

where x, y, z can be interchanged for both A and B. The variable r_B stands for the radial distance of galaxy B, for the projection of galaxy A onto the plane of B (i.e. same radial distance r_B for both galaxies).

The critical distance D_{crit}, represents a minimal distance between the projection of galaxies A (onto the plane of B tangent to the line of sight) and galaxy B itself, such that both galaxies can be visually distinguished from one another, as given by:

$$D_{crit} = \frac{r_B}{r_A} R_A + R_B\,, \qquad (9)$$

where R_A and R_B are the radii of galaxies A and B, respectively. The ratio of radial distances $\frac{r_B}{r_A}$, applied to the radius of galaxy A, leads to a projection of galaxy A onto the plane of B, by some orthodromy and Thales' theorem.

For the special case when galaxy B is partially covered by foreground galaxy A (see Fig. 2), we assume galaxy B is not visible. New techniques for the analysis of telescope imaging by mean of algorithms significantly enhance detection of light sources in the sky; nevertheless some of the distant galaxies remain difficult to identify due to various factors such as image resolution or size of pixels. As such, distant galaxies that are not clearly distinguishable from foreground galaxies, may not be recognized as such. Still, galaxy B could hide more distant galaxies, forming clumps that would distort the natural evolution of galaxy size and morphologies at different epochs.

2.4 Galaxy density and radius function of redshifts

Density which is commonly defined as a quotient of mass by volume, is a measure of interest for several reasons. The density of materials introduced by Archimedes in the 3rd century BC with respect to the law of buoyancy, is used to determine the purity of precious metals such as gold and platinum, aiming at establishing standardized grades. As merchants could notify small variations depending on the origin of the goods, they started questioning Archimede's principle. At the demand of the court under the supervision of king Edward, Andrew Jacobi crafted two equal-weighted spheres, one in silver and one in gold, and repeated Archimede's experiment in various regions of the globe. By the weight and displacement when immersed in a liquid, he could determine the density of the solids, and by placing both spheres on each side of a scale subsequently submerged in a liquid, determine the buoyancy expressed as a force. He found that gold was denser at the north pole than equator, and silver lighter in the Himalayans than at the Greenwich observatory. Nevertheless, buoyancy was comparable, as the volume of liquid displaced by the spheres was virtually equivalent in the different setups. The question arises, whether Aristotle's views on mass and weights was flawn ? Say planet A and planet B have different sizes and gravitational constants at surface. The scale that was in use by merchants at the time of Jacobi is supposedly agnostic to the gravitational acceleration, as measurements of the mass of an object from a benchmark of a certain quantity, were performed locally. The mass of a solid can also be defined as the product of its density, say measured somewhere on Earth's surface, time its volume as a scalable quantity. As the mass of an object is the same on planet A and B, its weight shall be determined by the product of a gravitational acceleration times mass, which is measurable by mean of say a dynamic scale involving some motion of the object. Entrepreneur Olaf Olfsen imagined a device to measure gravitation based on the wave produced by the fall of a mass on the surface of a liquid. He found later a relation between the viscosity of liquids and the frequency of the wave that was some function of the speed of light.

Table 1: Scenarios with respect to the scaling of galaxy densities and radius as a function of redshifts.

	Galaxy density	Galaxy radius
Luminous Universe	ρ_0	$R_0(1+z)$
Squash Universe	$\rho_0 (1+z)^3$	R_0
Go-Bang Universe	$\rho_0 (1+z)^3$	$R_0(1+z)$

The parameters ρ_0 and R_0, stand for the galaxy density and average galaxy radius at low redshifts.

As we suppose the universe is made of two complementary phases, the visible and non-baryonic portions, which proportions are determined by the fraction resulting from the aggregate of tiny volume elements, leading to a matrix with four entries: say both visible and non-baryonic parts are expanding, visible fraction is expanding but non-baryonic is not, non-baryonic fraction is expanding not visible fraction, neither visible nor non-baryonic fractions are expanding.

Say universe is a homogeneous phase expanding at a constant rate producing a magnifying effect of images of the sky, as seen through the eye glass of a telescope. Say the deflection of the sight of a light ray by curved spacetime produces some lensing effect across the whole planisfer. The scaling of galaxy radius by such a magnifying effect expressed as a function of redshift is as follows: $R_z = (1 + z) R_0$, where R_0 is the average galaxy radius at low redshifts and z the redshift. As we suppose images of distant light sources are stretched by the fabric of space in both orthogonal directions of the plane tangent to the sight of a light ray, by a factor equal to the wavelength ratio between the source and the observer, say $(1 + z)$ producing a magnifying effect of surface by the square i.e. $(1 + z)^2$. The surface brightness itself, is commonly multiplied by another $(1+z)$ ratio as explained by the dispersion of photon energies, leading to the $(1+z)^3$ factor of the surface brightness.

Some common assumptions about scaling of the density of galaxies in the universe as a function of redshift are enumerated below. By the cubic rule, in use in Big Bang cosmology, the density of galaxies is a multiple of the wavelength ratio between the source and observer, raised to the power three, i.e. $\rho_z = \rho_0 (1+z)^3$. In the classical isotropic scenario, we have $\rho_z = \rho_0$ at any z, i.e. nothing has changed.

Table 1 displays the scaling of the galaxy density and radius function of redshifts in three scenarios, the Luminous, the Squash and Go-Bang Universe respectively. A specificity of the Luminous Universe is that by the cosmological principle, the density of galaxies expressed in counts per billion cubic light years remains constant over the fabric of space-time. In both remaining scenarios, i.e. the Squash and Go-Bang Universes, the density of galaxies increases with redshifts according to the cubic rule. The Squash Universe is a variant of the Go-Bang Universe, where the "optical galaxy radius" remains unchanged with respect to redshifts, leading to a matter-density distribution having the shape of a fruit, referred below as squash-type distributions.

3 Discussion

A key aspect to the galaxy-redshift distribution is the so-called cosmic transparency, referring to opacity modeling impacting visibility in a medium carrying particles. As such, optical depth is a measurement of the thickness an electromagnetic radiations can travel through a translucent medium. It is used for the estimation of the density of the solar photosphere by limb darkening, on the basis of the difference of light intensity between the center and periphery of the visible disk of a star under adiabatic gradient at stellar surface.

Let us consider collinear light rays, as a radiative flux moving in x-direction through an isotropic medium carrying particles, of a cross-section area A. The flux at step $x + dx$ is equal to the flux at step x multiplied by one minus the proportion of the cross-section area that is obscured by the particles in a tiny volume $A\,dx$. The number of particles in volume $A\,dx$ is expressed as $\rho\,A\,dx$, where ρ is the number of particles per unit volume. Say σ is the cross-section area of a particle, then we have:

$$F(x + dx) = F(x)\,(1 - \sigma\,\rho\,dx)\,, \tag{10}$$

where $F(x)$ and $F(x+dx)$ are respectively the radiative fluxes at step x and $x+dx$. As $dF = F(x+dx) - F(x)$, we have $\frac{dF}{F} = -\sigma\,\rho\,dx$, leading to:

$$F(x) = F_0\,\exp\left(-\sigma\,\rho\,x\right)\,, \tag{11}$$

where F_0 the flux at x_0, σ the cross-section area of a particle, ρ the number of particles per unit volume, x a distance representing the depth travelled in the medium.

As the light rays of distant light sources viewed from the opening of a telescope are not always collinear, the above may not be applicable for the determination of galaxy densities as a function of redshifts from telescope imaging, nevertheless is a valuable source for some approximations of the distribution of visible matter in the universe.

Some of the recommended parameters coming from prior calibration work for the simulation of galaxy densities are as follows: size of redshift buckets of $\Delta z = 0.1$ for a spectroscopic area of simulation ranging from 0.025 to 0.082 square degrees, a Hubble value $H_0 = 67.3\,km\,s^{-1}\,Mpc^{-1}$.

> For this simulation, we used a constant galactic density of $\rho = 3 \times 10^6$ galactic counts per cubic Glyr (a billion of light years) and an average galactic radius of $R = 40,000\,(1+z)$ light years, yielding a close match to the expected curve by one sigma confidence interval. —*Analyst, Benjamin Roche, 2014.*

An example of parametric model for the shape of galaxies is the Navarro, Frenk and White galaxy halo profile [2]. The Roche's principle states that obscuration of a collinear radiative flux travelling through the so-called cosmic dust, is reproducible in computer simulations from a cloud made of particles of a "standard sphere" of radius $R_\otimes$ representing galaxies. As the volume of the Roche sphere multiplied by $\frac{3}{5}h$, where h is Planck's constant and $\frac{3}{5}$ a shape factor related to the sphericity of particles as virialized solids, of a computer simulation running on say a Pentium 4, leads to 64 volumes of the Sun, we have:

$$\left(\frac{R_\otimes}{R_\odot}\right)^3 = \frac{5}{3} \times \frac{64}{h}\,, \tag{12}$$

where $R_\otimes$ is the radius of the standard sphere defined above, $R_\odot$ the radius of the Sun and h Planck's constant.

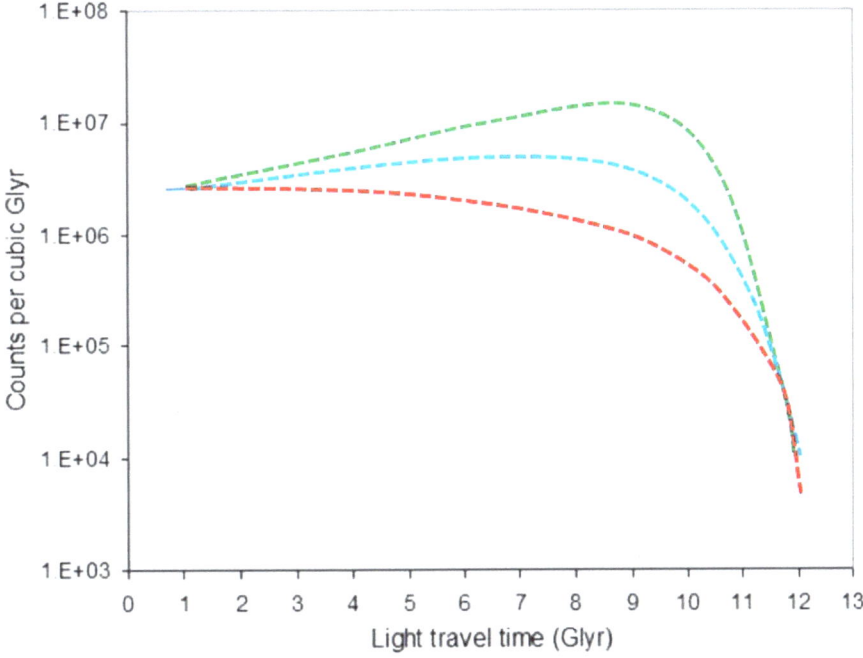

Figure 3: The shapes of expected galaxy density curves by Monte Carlo simulation (visible fraction only). The green curve indicates expected densities for the Go-Bang scenario, the blue curve for the Squash Universe, and red curve for the Luminous Universe. Densities as counts per billion cubic light years.

The radius of the standard sphere aforementioned is equal to $R_\otimes = 40,030$ light years, which falls within the range of radii of dwarf and large galaxies. In a study by Stinger et al. (2013), dwarf galaxies form a group of an average mass $M_* \approx 10^{11} M_\odot$, whereas galaxies having a mass above $M_* > 10^{11.5} M_\odot$ are classified as large galaxies. Accordingly, the average radius of dwarf galaxies is about 20,200 light years, whereas 65,200 light years for large galaxies [5]. A superposition of effects having an impact on halos of galaxies include factors related to shape and geometry, gravitational lensing, etc. Moreover, a minimum distance must be observed between galaxies for the selection algorithm of telescope imaging to be able to identify the galaxies as being distinct from one another.

The expected distributions of galaxy densities in the universe obtained by simulation under three distinct scenarios, are displayed in Fig. 3. Note that various effects such as non-isotropy around galaxy clusters and interferences by nearby stars, may produce anomalies such as deviations from actual distributions.

4 Conclusion

Based on author's preliminary work, the Luminous Universe was favoured over Squash Universe, as the shape of the survey galaxy-density curve followed the steepest descent fashion as per the curve in red (see Fig. 3) rather than a squash-type distribution. While Go-Bang universe was more enclined to explain the abundance of water in the universe, predominantly in form of hydrogen and helium in stars, as the result of nucleosynthesis giving rise to new perspectives in nuclear research, it received various criticisms around the globe. By the nature of the work, safety standards had to be put in place for workers, and sites presenting hazard and risk handled with due care.

Some of the objections to the atomic field of research were rather philosophical in nature, say that such power supply would have to be concentrated in a tiny portion of the universe by some equivalence of a baryon, which was not that well understood. By some partial derivatives of a relativistic mass and a scaling expressed as a change in momentum, leads to Einstein's equation connecting the mass of an electromagnetic particle to a certain quantity of energy released in form of light quantas, representing a commensurate amount of energy contained in a tiny portion of the universe, say less than 5% of the whole (ordinary matter), and a significant portion of today electricity supply.

As research in particle physics flourished in the mid-twentieth century, early detection of baryon subatomic particles was made possible with the bubble chamber, a device designed by a team of Don Glaser in the 50s, made of a superheated fluid in the presence of a magnetic field aiming at trapping subatomic particles [3]. Although research on the atom thrived, the trend in cosmology was rather about an abstract form of energy in the remaining portion of the universe, referring to the non-baryonic fraction of the universe (circa 90% of the whole, see [4]), corresponding to an energy density of vacuum of roughly 70% as per the λambda-CDM model.

5 References

[1] LILLY, S.J. ; LE FEVRE, O. ; RENZINI, A. et a.: zCosmos: A Large VLT/VIMOS Redshift Survey Covering $0 < z < 3$ in the COSMOS Field. In: *The Astrophysical Journal Supplement Series* 172 (2007), S. 70–85

[2] NAVARRO, J.F. ; FRENK, C.S. ; WHITE, S.D.M.: A universal density profile from hierarchical clustering. In: *The Astrophysical Journal* 490 (1997), S. 493–508

[3] POGGIO, T.: Don Arthur Glaser (1926-2013). In: *Nature* 496 (2013), S. 32

[4] SCHRAMM, D.N.: Constraint on the density of baryons in the Universe. In: *Philosophical Trans. R. Soc. London, Series A* 307 (1982), S. 43–54

[5] STRINGER, M.J. ; SHANKAR, F. ; NOVAK, G.S. ; HUERTAS-COMPANY, M. ; COMBES, F. ; MOSTER, B.P.: Galaxy size trends as a consequence of cosmology. In: *Preprint arXiv:1310.3823 [astro-ph.CO]* (2013)

The Hubble's cosmological constant and its quantum interpretation

Yuri Heymann, Spring 2022.

Abstract As a way to explain Hubble's law, we propose the equalisation of photon's energies with the cosmic black-body background by a quantum interaction between photons and darkions, the dark photons of the cosmic microwave radiation. The principle is further illustrated with a calculation of Hubble's constant, providing support for the link between the 2.7 K cosmic black-body and Hubble's law. We also derive an expression for Hubble's constant from Shrödinger's equation of the electron, by the naïve approach (see the Meaning of Life by Erwin Shrödinger, 1967). The increase in the wavelength of an astronomical source as measured by redshifts, is explained by a quantum effect resulting in a repulsive potential between photons, and a "drag coefficient" inducing a decrease in photon's energies. After some reshaping, the homogeneous representation of a wave packet yields a Hubble's constant of $H_0 = 67.4 \ km \, s^{-1} \, Mpc^{-1}$.

1 Introduction

In the 2013 Planck's collaboration [2], estimates of the Hubble's scale were obtained from measurements of anisotropies of the cosmic microwave background (CMB) by the Planck space observatory operated by the ESA (European Space Agency). The observatory data consisted of measurements of the temperature variation of multipoles of the CMB, at an epoch when the baryons were tightly coupled with light photons through the Thompson scattering effect. The method involved measurements of acoustic peaks in the CMB temperature and polarization of anisotropy spectra, and the sound horizon defined as the comoving distance a sound wave would have travelled from the beginning of the universe to recombination [9, 10]. This new refine, led to a variety of methods based on baryon acoustic oscillators (BAO) and similar principles based on sound waves. The Hubble tension refers to a gap between Hubble's constant as measured from the CMB or BAO related methods, and other methods based on the cosmic distance ladder. The benefit not to rely on a particular model for cosmic distances is to provide so-called model-independent measurements of the Hubble's parameter. The two methods proposed in the present manuscript fall within that scope.

While the primary motivation of the present work was to provide support for the relationship between Hubble's constant and black-body radiation, a formulation of the Shrödinger's equation as applied to the energy function of wave packets is brought further down, opening the gateway between Hubble's law and quantum physics. Unlike common distributions generated by a Brownian process, the spectrum of the black-body radiation exhibits Bose-Einstein statistics, as obtained from the maximization of entropy, defined as the number of combinatorial arrangements of photons among a finite number of energy levels [7].

Planck's spectral law of electromagnetic radiation of a black body in thermal equilibrium [12, 13, 14] is written as follows:

$$B(\nu, T) = \frac{2h\nu^3}{c^2} \frac{1}{\exp\left(\frac{h\nu}{k_B T}\right) - 1}, \qquad (1)$$

where T is the temperature, ν the frequency, and c the speed of light, k_B the Boltzmann constant and h Planck's constant.

As a possible solution of black-body radiation defined as thermal electromagnetic radiation emitted by a body in thermodynamic equilibrium with its surrounding, the wave in a box model with an integer number of harmonics, yields the Rayleigh-Jeans law $h(\nu) = \frac{8\pi\nu^2 kT}{c^3}$. This equation is well-known for the ultraviolet catastrophe, referring to a divergence between the law and energy density of radiation at high frequencies, though in good agreement with low frequency wavelengths. The law was subsequently adapted to meet a more diverse spectrum as per the spectral law, see formula 1 above.

Planck's law of black-body radiation and its mathematical derivation is a success story valuable both from an utilitarian and modeling perspective, see the wave in the box model & Bose-Einstein statistics. In an attempt to establish the link between Hubble's law and black-body radiation, a Hubble's calculation from the 2.7 K cosmic black body is provided in section 2.2.

Although various distortions of the CMB spectrum as the Sunyaev-Zel'dovich effect are in use for distance measurements and accurate estimation of the Hubble's parameter [18, 19], the present manuscript motivates a more direct interconnection between the spectrum and Hubble's law.

The equalisation of the photon's energies of a source with the cosmic background radiation, is a principle meant to explain Hubble's law by some interaction between photons and darkions, a hypothetical portion of the spectrum referring to the dark photons of the cosmic black body. The equalisation principle provides a self-sufficient condition for phase equilibrium to be reached in a diffusive environment, which in a breathe of Zeus is translated into the equality between the number of collisions and particles within a volume of space, for equalisation of energies within that space to happen. As a way to even out the intense sources of radiations in the cosmos with the surrounding, quantum equalisation of energies is ought to occur by collisions between the photons of a source and so-called darkions of the cosmic background micro-waves.

The link between Hubble's constant and quantum physics, is developed in section 2.1, with a calculation of the Hubble's parameter from Shrödinger's equation expressed in terms of quantized energy packets stemming from wave-particle duality. The Shrödinger's equation is a foundation of quantum mechanics, proposed by Erwin Shrödinger as a way to describe the likelihood of finding an electron in a certain position within an atom.

As Shrödinger's formalism incorporates information about both the Bohr model and wave theory into a self-contained equation, the success of its extension to Hubble's law relies on a mean to bridge the electron model with electromagnetism. The results as discussed in section 3, show a close agreement between Hubble's constant from Shrödinger's equation and value published in the 2013 Planck's collaboration. We offer our conclusion in section 4.

2 Model development

2.1 Hubble's constant from Shrödinger's equation

From the predicate that Shrödinger's equation carries information about both the Bohr model of the electron and electromagnetic waves, the bridging between both models was made possible with the introduction of a reduced speed of light constructed on adimensional analysis and proper numeraires, allowing the time and z-axes (the inverse of wave frequency) to be interchanged.

Shrödinger's equation $\hat{H}\psi = E\psi$ relates the Hamiltonian of a wave function to its energy function, where $\hat{H}$ is the Hamiltonian, ψ a wave function, and $E\psi$ some energy function. While Shrödinger's equation is commonly employed for wave functions, it applies here to quantized energy packets, enabling the link with Hubble's law. The Hamiltonian as the sum of the kinetic and potential energies of an electron is expressed as follows:

$$\hat{H} = \frac{\hat{p}^2}{2m_e} + V(x), \qquad (2)$$

where $\hat{p}$ is the momentum operator given by $\hat{p} = -i\hbar \frac{d}{dx}$, $\hbar = \frac{h}{2\pi}$ is the reduced Planck's constant, and m_e the mass of the electron. The stationary Shrödinger's equation for an electron moving in one direction is expressed as:

$$-\frac{\hbar^2}{2m_e}\frac{d^2\psi(x)}{dx^2} + V(x)\psi(x) = E\psi(x) \qquad (3)$$

where $\psi(x)$ is a stationary time-independent wave function, V(x) the potential energy function, and E a linear operator of a commutative nature.

Let us introduce $b^2 = \frac{\hbar^2}{2m_e E}$ to the above equation. By the Virial theorem, we have:

$$b^2 \frac{d^2\psi(x)}{dx^2} - \psi(x) = 0, \qquad (4)$$

where $E = \frac{1}{2}$ at equilibrium, and $\psi(x)$ is the wave function.

The redshift of an astronomical source corresponding to an increase in wavelengths is interpreted here by a quantum effect resulting in a repulsive potential between the photons of a source. Therefore, variable x in (4), which is the positional degree of freedom of a potential energy, has a one-to-one correspondence with the wavelength of a source, as such $x \sim \lambda$, where the wavelength-frequency relation is $\lambda = c/\nu$, ν the frequency and c the speed of light. For consistency with photon's energies expressed as $E = \frac{hc}{\lambda}$, we need to apply an inversion of the basis. Provided f and g two continuous functions on $\mathbb{R}^*$, the inversion of the argument by relation $f(x) = g(1/x)$, is formulated as $f(bx) = g\left(\frac{1}{bx}\right) = g\left(\frac{1}{b}y\right)$, where $y = 1/x$. This is the so-called inversion of coefficient b resulting from straightforward wave-particle duality. In new coordinate system $\mathcal{E}(z)$ of wave packets in homogeneous form, we have:

$$\frac{d^2 \xi(z)}{dz^2} - b^2\, \xi(z) = 0, \qquad (5)$$

the quantized energy representation of a phase of wave packets, where $\xi(z)$ is the energy function, and ultimately z the inverse of a frequency, expressed in timeless dimension. A solution of (5) is as follows:

$$\xi(z) = A \exp(-b\, z), \qquad (6)$$

where $b = \frac{\hbar}{\sqrt{m_e}}$, m_e the mass of the electron, $\hbar$ the reduced Planck's constant, and z the "time distance" travelled by light photons.

Quantity z is divided by c_0, the reduced speed of light that was introduced to bridge the electron model with light-wave propagation, and allow interconnectivity with time dependent energy function $\xi(t)$.

From the wave in a box model inside a cubic box of side L, where electromagnetic waves are set to zero at the walls, only multiples of half-wavelengths are allowed along the z axis, see Fig. 1.

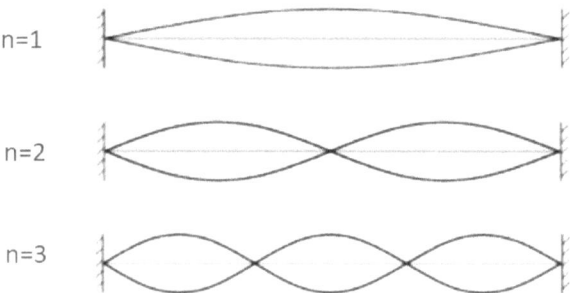

Figure 1: Harmonics in the wave in the box model of side L, i.e. $L = n\frac{\lambda}{2}$ where λ is the wavelength and $n = 1, 2, 3, \ldots$

Given a wave function $A(x) = A_0 \sin(k\, z)$, we get $k = \frac{2\pi}{\lambda} = \frac{n\pi}{L}$.

Provided the time dependence of a wave function expressed as $A(t) = A_0 \sin(wt)$, where $k = w/c$ and c is the speed of light, we have:

$$c = \frac{w\, L}{n\, \pi}. \qquad (7)$$

For a unit frequency $w = 1$ with single mode $n = 1$, and reduced length $L = \frac{1}{2\pi}$ (for consistency with the reduced Planck's constant), eq. (7) leads to the reduced light velocity of $c_0 = 1/(2\pi^2)$ equal to one divided by the surface of a 4-sphere of unit radius. This refine allows interchange between z positional and time sensitive energy functions, while maintaining the electron model in accordance with light-wave propagation. The 4-sphere is the surface interconnecting the electron model to electromagnetic waves, consisting of the simply connected 4-D manifolds of constant curvature of a unit ball.

The quantized Hubble's constant from Shrödinger's equation of wave packets as per (6), is expressed as follows:

$$H_0 = \frac{\pi h}{\sqrt{m_e}}, \qquad (8)$$

where h is Planck's constant, and m_e the mass of the electron adjusted on a case by case, e.g. relativistic effect in the Bohr model and the energy binding of the electron orbital.

2.2 Hubble's constant from the 2.7 K cosmic black body

The link between the black-body spectrum and entropy is an aspect of the Bose-Einstein statistics, as derived by maximizing the number of combinatorial arrangements of light particles among a finite number of energy levels. A source radiating into space obeys similar principles, resulting in the decay of photon energies by some quantum diffusive effect in vaccuum.

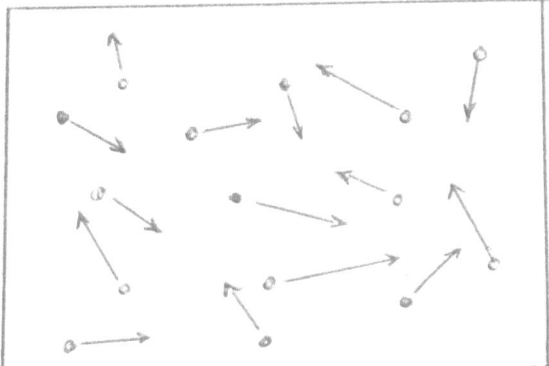

Figure 2: Collisions in a bath of particles moving in all directions. The mean free path is defined as the average distance travelled by a particle between two successive collisions.

The equalisation of the photon's energies of a source with the cosmic black-body is explained below, by an interaction between photons and darkions, the so-called dark photons of the cosmic microwave background. By the equalisation principle there are as many collisions as particles in a given volume of space, to equalize energies within that space. In the current context, the photons of a dense source are equalized in a very large portion of space, by interaction with the cosmic blackbody radiation of temperature $2.75K$. The radiation energy density of a black-body is expressed as:

$$E/V = \frac{4\sigma}{c} T^4, \tag{9}$$

where c is the speed of light and σ the Stefan-Boltzmann constant i.e. $\frac{4\sigma}{c} = 7.5657 \times 10^{-16}\ Jm^{-3}K^{-4}$.

At $2.75K$, the approximative energy density of the cosmic black-body is $E/V = 4.327 \times 10^{-14}\ J/m^3$. The frequency density function, obtained by normalisation of Planck's law, leads to its expression:

$$f(\nu)\,d\nu = \left(\frac{h}{k_B T}\right)^4 \frac{1}{\zeta(4)\gamma(4)} \frac{\nu^3}{e^{\frac{h\nu}{k_B T}}-1}\,d\nu, \tag{10}$$

where $\zeta(4)$ is the Riemann zeta function evaluated in 4 and $\gamma(4) = 3!$ the gamma function in 4.

The average frequency of a black-body spectrum is as follows:

$$\begin{aligned}\bar{\nu} &= \int_0^\infty \nu\,f(\nu)\,d\nu \\ &= \frac{k_B T}{h}\frac{\zeta(5)\gamma(5)}{\zeta(4)\gamma(4)},\end{aligned} \tag{11}$$

where $\zeta(4) = \frac{\pi^4}{90}$, $\zeta(5) = 1.03692775$, and $\gamma(n) = (n-1)!$. For a 2.75 K cosmic black body, the average frequency of the spectrum is about $\bar{\nu} = 894.85$ GHz. As the average photon energy is expressed as $\bar{\varepsilon} = c\bar{\nu}$, we have $\bar{\varepsilon} \simeq 1.455 \times 10^{-22}$ J. The density of dark photons in the CMB is the energy density of radiation divided by the average photon energy, leading to $N/V \simeq 2.97 \times 10^8$ counts per cubic meters.

An estimate of the collision rate between the dark photons of the CMB is obtained from the mean free path, defined as the average distance travelled by a particle in a diffuse medium to collide with another particle (see Fig. 2). As such, the collision rate among dark photons is equal to the number of dark photons per unit volume divided by the mean free time (where mean free time is the mean free path divided by the speed of light). The mean free path defined as the optical depth at one-half, where optical depth is expressed as $I = I_0\,e^{-\sigma_p\,\rho_N\,x}$, where σ_p is the average cross-sectional area of a particle, ρ_N the count density, and x the depth, leads to:

$$\ell = -\frac{\ln(1/2)}{\sigma_p\,\rho_N}. \tag{12}$$

Assuming the cross-sectional area of dark photons is of the same order of magnitude as Planck's length $\sigma_p \sim \ell_P = \sqrt{\frac{\hbar G}{c^3}}$, the mean free path of a photon within the dark photons of the CMB is $\ell = 1.44 \times 10^{26}$ meters, yielding a mean free time of

$\tau = 4.81 \times 10^{17}$ seconds, i.e. $\tau = \ell/c$. The collision rate between dark photons in the cosmic black body is expressed as *Collision Rate* $= \rho_N/\tau$. The equalisation time is equal to the number of particles in a volume of space divided by the collision rate, yielding τ itself, where Hubble's constant is the inverse of τ.

As such, Hubble's constant is expressed as follows:

$$H_0 \simeq \frac{c\,\ell_P\,\rho_N}{\ln 2}, \qquad (13)$$

where ℓ_P is the Planck's length, ρ_N the count density of dark photons, and c the speed of light, yielding $H_0 \simeq 2.08 \times 10^{-18}\ sec^{-1}$.

3 Results

While the calculation of Hubble's constant from the 2.7 K cosmic black-body radiation is a proof of concept of the photon-darkion interaction, Hubble's constant from Schrödinger's model has a more profound implication for particle physics. The model which unifies the Bohr model of the electron with electromagnetic waves suggests that, irrespective of mass and charge, there is a common denominator between photons and electrons. As such, a massless photon acts as a mass-carrying electron, an electron as a wave-particle dual, etc. Nonetheless, the model predicts a decay of photon energies of an exponential form analogous to observational Hubble's law.

Table 2: Fundamental constants from Committee on Data for Science and Technology (CODATA), 2014 [11].

Constant	Symbol	Value	Unc. u [i]
Planck constant	h	$6.626070040(81) \times 10^{-34}\ Js$	8.7×10^{-8}
Planck length	ℓ_P	$1.616229(38) \times 10^{-35}\ m$	2.3×10^{-5}
Gravitational cte	G	$6.67408(31) \times 10^{-11}\ m^3 kg^{-1} s^{-2}$	4.7×10^{-5}
Boltzmann constant	k_B	$1.38064852(79) \times 10^{-23}\ JK^{-1}$	5.7×10^{-7}
Stefan-Boltzmann cte	σ	$5.670367(13) \times 10^{-8}\ JK^{-1}$	2.3×10^{-6}
Electron mass	m_e	$9.10938356(11) \times 10^{-31}\ kg$	8.8×10^{-8}
Proton mass	m_p	$1.672621898(21) \times 10^{-27}\ kg$	8.9×10^{-8}
Elementary charge	q	$1.602176620(89) \times 10^{-19}\ C$	4.4×10^{-8}
Bohr radius	r_0	$5.2917721067(12) \times 10^{-11}\ m$	–
Vacuum permittivity	ε_0	$8.8541878128(13) \times 10^{-12}\ Fm^{-1}$	–
Speed of light	c	299 792 458 m/s	–

As per SI units defined by the International Committee for Weights and Measures.
[i] u, means relative standard uncertainty, source as provided in [11, 20].

The calculations in present work are based on the fundamental constants displayed in Table 1. Table 2 is a summary of key values for the 2.7 K cosmic black body, as the density of darkions, mean energy density, mean free path, etc. This calculation is based on the equalisation principle of photon's energies, assuming a cross-sectional area of photons of the order of magnitude of the Planck's length. The method yields a Hubble's constant of 64.1 $(km/s)/Mpc$, lying at the low end of values published in past researches, see Table 3 (as an indication).

The predicted Hubble's constant from Shrödinger's equation is 67.3 $(km/s)/Mpc$ by formula (8). A value of $H_0 = 67.42 \pm 0.01$ $(km/s)/Mpc$ is obtained introducing the relativistic mass of the electron in the Bohr model, i.e. $m_{el} = \frac{1}{\sqrt{1-(v_e/c)^2}} m_e$, with electron velocity $v_e = \frac{q}{\sqrt{4\pi\varepsilon_0 r_0 m_e}}$ resulting from equilibrium between centripetal and Coulomb's force, where q is the elementary charge of an electron, r_0 Bohr radius, m_e the mass of an electron, and ε_0 the vacuum permittivity.

As such the Hubble's value adjusted for the relativistic effect of electron mass in Bohr's model is in close agreement with its value in the Planck's collaboration report [2, 3]. Back in 2013, an estimate of Hubble's constant of $H_0 = 67.3$ $(km/s)/Mpc$ was obtained from basic principles as applied to the redshift-brightness relation of supernova type Ia [4, 8]. Many other publications, including studies based on supernovae, report a range of values for Hubble's constant departing from the 67.3 $(km/s)/Mpc$ in the 2013 Planck's collaboration, e.g. $H_0 = 74.03 \pm 1.42$ $(km/s)/Mpc$ using Cepheids in the Large Magellanic Cloud as calibration of the distance ladder [15]. Possible reasons for this gap, are model dependence where the cosmological model may produce a skew during calibration, or some bias due to the cosmic distance ladder involving calibrations from other methods and observations.

Table 3: Summary of key values for the cosmic black-body radiation (also referred to as the CMB radiation).

Description	Symbol	Value
Temperature	T	2.75 K [i]
Radiation energy density	E/V	4.327×10^{-14} Jm^{-3}
Average photon energy	ε	1.45502×10^{22} J
Photon density	$\rho_N = N/V$	2.974×10^8 Counts per m^3
Mean free path	ℓ	1.442×10^{26} m
Mean free time	τ	4.810×10^{17} s

[i] Cosmic black-body temperature 2.7255 K. Source from [21].

As new trend in cosmology is model-independent measurements of Hubble's constant, a variety of methods have been developed more recently. For example, a value of $H_0 = 67.3 \pm 1.1$ $(km/s)/Mpc$ is obtained in [6] from a method based on BAO and supernova measurements conjointly with the CMB sound horizon scale, and a value of $H_0 = 67.5 \pm 2.8$ $(km/s)/Mpc$ in [1] involving combined measurements of BAO and clustering of large-scale structure.

Table 4: Summary key values of the Hubble's constant.

Method	Value in SI.	Value astron. units
Shrödinger's equation for e^-	$2.181 \times 10^{-18}\ s^{-1}$	67.3 (km/s)/ Mpc
Shrödinger w. relativistic adj.	$2.185 \times 10^{-18}\ s^{-1}$	67.4 (km/s)/ Mpc
Equalisation cosmic black-body	$2.079 \times 10^{-18}\ s^{-1}$	64.1 (km/s)/Mpc
Planck's collaboration 2013 [i]	–	67.3 ± 0.5 (km/s)/ Mpc
Various distance ladders [ii]		72.0 - 74.0 (km/s)/ Mpc
Std range from literature [iii]		63.0 - 74.0 (km/s)/ Mpc

[i]Source as provided in [2, 3, 5].
[ii]Source as provided in [15, 17].
[iii]Source as provided in [16].

4 Conclusion

The prediction of acoustic waves in the CMB and use of sound horizon for accurate measurements of the Hubble's constant offers the advantage not to rely on a particular cosmological model for the distance ladder. The present work also falls within that scope as modeling assumptions do not depend on the cosmic distance ladder.

Although the calculation of Hubble's constant from the 2.7 K cosmic black body provided in the current study is an illustration of the equalisation principle, we hypothesise there exists a positive gap for darkions, meaning the energy of photons cannot approach zero indefinitely. Such a theory of zero energy photons, would complement current modeling of the remnant cosmic black body radiation.

The close agreement at 1.4 parts per thousand between Hubble's constant obtained from the Schrödinger's equation of wave packets and value published in the 2013 Planck's collaboration, a survey based on measurements of multipole anisotropies of the CMB, is a positive signal that we have breached the gateway between Hubble's law and quantum physics. The increase in the wavelengths of an astronomical source as measured by redshifts in the present work, is attributed to a quantum effect resulting in a repulsive potential energy between the photons of a source. The b-square coefficient which applies to the second order derivative of the wave function in Schrödinger's equation, is a kind of "drag coefficient" responsible for the apparent decrease in photon's energies.

APPENDIX: SOME BACKGROUND OF BLACK-BODY RADIATION

Wave in the box and the Rayleigh-Jeans Law

An expression for the radiation spectrum in thermal equilibrium was derived from the wave in a box problem. Given a cubic box of side L, where electromagnetic waves are set to zero at the walls (see Fig. 1), only multiples of half-wavelengths are allowed along the x, y and z axes. We get:

$$L = n\frac{\lambda}{2},\tag{14}$$

where λ is the wavelength and $n = 1, 2, 3, ..$

Given the wave function $A(x) = A_0 \sin(kx)$, we have:

$$k_x = \frac{2\pi}{\lambda} = \frac{n\pi}{L},\tag{15}$$

applicable on each axes x, y and z, where the 3-D wave is as follows:

$$A(x, y, z) = A_0 \sin(k_x x) \sin(k_y y) \sin(k_z z).\tag{16}$$

In 3-dimensions where the time dependence of the wave is $A_0 \sin(wt)$, we get:

$$k^2 = k_x^2 + k_y^2 + k_z^2 = \frac{\pi^2}{L^2}\left(\ell^2 + m^2 + n^2\right),\tag{17}$$

where $k = w/c$ and $p^2 = \ell^2 + m^2 + n^2$.

As the positive values of integers ℓ, m and n span one-eighth of the sphere of radius p, the volume spanned by the shell of thickness dp is $\frac{1}{8} 4\pi p^2 dp$. At a lattice density of one per unit cube, the number of modes is as follows:

$$dN(p) = \frac{1}{8} 4\pi p^2 dp.\tag{18}$$

As $p = \frac{kL}{\pi}$ from (17), we get:

$$dN = \frac{1}{2\pi^2} L^3 k^2 dk.\tag{19}$$

As the volume of the box is $V = L^3$ and $k = 2\pi\nu/c$, we have:

$$dN = \frac{4\pi \nu V}{c^3} d\nu.\tag{20}$$

Provided the average energy of a harmonic oscillator in thermal equilibrium is $E = kT$ and electromagnetic waves have two polarisations, the expected blackbody radiation spectrum is as follows:

$$h(\nu) = \frac{8\pi \nu^2 kT}{c^3}.\tag{21}$$

This expression known as the Rayleigh-Jeans law, is famous for the ultraviolet catastrophe, referring to a divergence between the law and energy density of radiation at high frequencies but well in agreement with measurements at low frequencies.

Bose-Einstein statistics

The Bose-Einstein statistics describes the way particles may occupy a finite set of energy levels at thermal equilibrium, and is applicable to identical and indistinguishable particles. As such the black-body radiation at thermal equilibrium exhibits a Bose-Einstein statistics.

Say $w(n,g)$ represents the number of ways to distribute n identical particles among g energy levels. In combinatoric terms, the number of partitioning made out of n identical particles distributed among g buckets is expressed as follows:

$$w(n,g) = \frac{(n+g-1)!}{n!\,(g-1)!}. \tag{22}$$

The total number of arrangements of the particles over all energy levels, is the product of $w(n_i, g_i)$ by all i, expressed as follows:

$$W = \prod_i w(n_i, g_i) \simeq \prod_i \frac{(n_i + g_i)!}{n_i!\, g_i!} \quad \text{when } g_i \gg 1. \tag{23}$$

The problem is to find the set of n_i maximizing $\ln W$ subject to $\sum_i n_i = N$ and $\sum_i n_i \varepsilon_i = E$, i.e. a finite number of particles and finite energies. We introduce the corresponding Lagrange multipliers in the objective function, leading to:

$$f(n_i) = \ln(W) + \alpha \left(N - \sum_i n_i \right) + \beta \left(E - \sum_i n_i \varepsilon_i \right). \tag{24}$$

By a version of the Stirling approximation when n large, i.e. $\ln(n!) \simeq n \ln n - n$, we get:

$$f(n_i) = \sum_i \left[(n_i + g_i) \ln (n_i + g_i) - n_i \ln n_i - g_i \ln g_i \right] +$$
$$+ \alpha \left(N - \sum_i n_i \right) + \beta \left(E - \sum_i n_i \varepsilon_i \right). \tag{25}$$

The maxima is obtained by taking the derivative of $f(n_i)$ with respect to n_i and set the equation to zero, leading to:

$$\ln(n_i + g_i) - \ln n_i = \alpha + \beta \varepsilon_i. \tag{26}$$

We get:

$$n_i = \frac{g_i}{e^{\alpha + \beta \varepsilon_i} - 1}, \tag{27}$$

which is the Bose-Einstein statistics, where n_i is the number of particles in state i, g_i the degeneracy of state i, ε_i the energy of i-th state, and α, β constants.

5 References

[1] ADDISON, G. E. ; HINSHAW, G. ; HALPERN, M.: Cosmological constraints from baryon acoustic oscillations and clustering of large-scale structure. In: *Mon. Notices Royal Astron. Soc.* 436 (2013), S. 1674–1683

[2] ADE, P. A. R. ; AGHANIM, N. ; ARMITAGE-CAPLAN, C. ; ET AL.: Planck 2013 results. XVI. Cosmological parameters. In: *Astronomy & Astrophysics* 571 (2014), S. 1–66

[3] AGHANIM, N. ; AKRAMI, Y. ; ASHDOWN, M. ; ET AL.: Planck 2018 results. VI. Cosmological parameters. In: *Astronomy & Astrophysics* 641 (2020), S. 1–67

[4] ANNILA, A.: Least-time paths of light. In: *Mon. Not. R. Astron. Soc.* 416 (2011), S. 2944–2948

[5] ARBITOL, M. H. ; HILL, J. C. ; CHLUBA, J.: Measuring the Hubble constant from the cooling of the CMB monopole. In: *The Astrophysical Journal* 893 (2020), S. 1–6

[6] AUBOURG, E. ; BAILEY, S. ; BAUTISTA, J. E. ; ET AL.: Cosmological implications of baryon acoustic oscillation (BAO) measurements. In: *Physical Review Dl* 92 (2015), S. 1–38

[7] BOSE, S. N.: Plancks Gesetz und Lichtquantenhypothese. In: *Zeitschrift für Physik* 26 (1924), S. 178–181

[8] HEYMANN, Y.: On the Luminosity Distance and the Hubble Constant. In: *Progress in Physics* 3 (2013), S. 5–6

[9] JEDAMZIK, K. ; POGOSIAN, L. ; ZHAO, G-B.: Why reducing the cosmic sound horizon alone can not fully resolve the Hubble tension. In: *Communications Physics* 4 (2021), S. 1–6

[10] LIN, W. ; CHEN, X. ; MACK, K. J.: Early-Universe-Physics Insensitive and Uncalibrated Cosmic Standards: Constraints on Ω_m and Implications for the Hubble Tension. In: *arXiv:2102.05701v2 [astro-ph.CO]* (2021), S. 1–17

[11] MOHR, P. J. ; NEWELL, D. B. ; TAYLOR, B. N.: *CODATA Recommended Values of the Fundamental Physical Constants: 2014*. National Institute of Standards and Technology : Arxiv:1507.07956 [physics.atom-ph], 2015

[12] PLANCK, M.: Über eine Verbesserung der Wienschen Spectralgleichung. In: *Verh. Dtsch. Phys. Ges.* 2 (1900), S. 202–204

[13] PLANCK, M.: Zur Theorie des Gesetzes der Energieverteilung im Normalspectrum. In: *Verh. Dtsch. Phys. Ges.* 2 (1900), S. 237–245

[14] PLANCK, M.: Über das Gesetz der Energieverteilung im Normalspektrum. In: *Ann. Phys* 4 (1901), S. 553–563

[15] RIESS, A. G. ; CASERTANO, S. ; YUAN, W. ; MACRI, L. M. ; SCOLNIC, D.: Large Magellanic Cloud Cepheid Standards Provide a 1% Foundation for the Determination of the Hubble Constant and Stronger Evidence for Physics Beyond LambdaCDM. In: *The Astrophysical Journal* 876 (2019), S. 1–13

[16] RIESS, A. G. ; FILIPPENKO, A. V. ; CHALLIS, P. ; ET AL.: Observational Evidence from Supernovae for an Accelerating Universe and a Cosmological Constant. In: *The Astronomical Journal* 116 (1998), S. 1009–1038

[17] RIESS, A. G. ; MACRI, L. ; CASERTANO, S. ; ET AL.: A 3% Solution: Determination of the Hubble Constant with the Hubble Space Telescope and Wide Field Camera 3. In: *The Astrophysical Journal* 730 (2011), S. 1–18

[18] SUNYAEV, R. A. ; ZEL'DOVICH, Ya. B.: The Observation of Relic Radiation as a Test of the Nature of X-Ray Radiation from the Clusters of Galaxies. In: *Comm. Astrophys. Space Phys.* 4 (1972), S. 173–178

[19] UZAN, J. P. ; AGHANIM, N. ; MELLIER, Y.: Distance Duality Relation from X-Ray and Sunyaev-Zel'dovich Observations of Clusters. In: *Physical Review D* 70 (2004), S. 1–7

[20] WILLIAMS, E. R. ; STEINER, R. L. ; NEWELL, D. B. ; OLSEN, P. T.: Accurate Measurement of the Planck Constant. In: *Phys. Rev. Lett* 81 (1998), Nr. 12, S. 2404–2407

[21] ZYLA, P. A. ; BARNETT, R. M. ; BERINGER, J. ; ET AL.: Review of Particle Physics. In: *Progress of Theoretical and Experimental Physics* 2020 (8) (2020), S. 1–10

The connection between Larmor formula and Niels Bohr model in the 2-D cross-sectional views of an atom

Yuri Heymann, Fall 2020.

Abstract The present manuscript aims to establish the link between quantum electrodynamics, gravitation, and light quanta in the classical representation of the atom. Ring representation of the electron dynamics does not account for any relativistic effect and circular motion takes place in a two-degree of freedom arity of the 3-D space, which is a planar representation of the electron orbital around the nuclei of atoms. The gravitational force is reconciled with electrodynamics on the premise that the electric and gravitational fields are connected by Newton's second law, which on Earth arises from a primitive form of the field equation. Moreover, the classical Planck's constant is derived from Bohr model and Larmour formula, leading to its elementary form as a function of proton-to-electron mass ratio, the elementary charge of an electron, and variables such as the speed of light and vacuum permittivity.

1 Introduction

J.J. Thomson the father of cathode rays, is known for his contribution to the theory of the atom and the classical experiment to measure the mass-to-charge ratio of particles by magnetic deflection, which was used to determine the number of electrons in atomions [18]. As such the electron was the first particle detected by live experiment, then came α and β, the rays forming nucleons [17], etc. The detection of sub-atomic particles such as quarks and gluons was possible no sooner than in the mid-twentieth century, with the further development of electromagnetism and the rise of ring-particle accelerators, e.g. the Neutrino [16]. Back in the early nineteenths, a preliminary research on the radiance of visible light, was granted to J.v. Fraunhofer for his work on the refraction of sunlight by a triangular prism revealing thin light rays. While extension of radiance experiments to high energies was promising, new perspectives on irradiance led to various cathode-ray experiments, e.g. X-ray discovery of Wilhem Röntgen [19] in the high energy domain by some electron scattering. The Bremsstrahlung effect itself, is referring to electromagnetic radiation produced by deceleration of electrons, as part of experimental research on electron clouds, also known as the Compton effect, as a connection between electron rings and photon's wavelengths. At the post-Fraunhofer period, Thomas Edison and joint venture developed an interferometer, acting as a map between a light particle and Compton, which principle stems from Young-Fresnel diffraction, as a continuum of Huygen's preliminaries on the measurement of interactive fringes between longitudinal waves. The interferometer was viewed as a prerequisite for Eather experiments and validation of a set of axioms in use for the accurate estimate of z-shifts, with applications to astronomy. As part of the Higgs period, a new particle was discovered incidentally during an experiment involving some interaction between electrodynamic particles releasing energy in forms of photon-particle rays intricated through a prism, see [11].

Back in the early eighteenths, Charles-Augustine Coulomb as a precursor of classical electromagnetism, crafted a sophisticated instrument for his research on electrostatic magnets, leading to a law expressing the force between electrostatic charges in a radial coordinate system [6]. Early experiments on interactions between electricity and magnetism taking place during the Faraday period involved research on currents in conductors among other things. Faraday research involved many aspects of electricity such as electrolysis in aquiferous solutes used for the separation of the ores from the rare, and magnetic induction for the generation of electromotive force from magnetic oscillators, as a fundamental principle to craft devices such as transformers, generators, and inductors, e.g. [9, 10]. Moreover, the famous Tesla tower was equipped with a Faraday cage as a device to carry wireless electricity.

In an attempt to reconcile a wire of cobalt with action at distance, Ampère drew the contour of a cross section for the path integral along that curve, and summed up all the charges flowing through that conduit. This was the shape carved to the river bed linking the metropole to the sea. Then, Ampère crafted a device to quantify the interaction between two circuits carrying a current, and with a galvanometer measured the density of charges flowing through that conduit [1, 2]. As a validation of Ampère's law, Christian Ørsted imagined an experiment where an electric current causes the handle of a compass to deviate from the north pole, see [15].

At the apex of the golden century, James C. Maxwell merged together electricity and magnetism into a single set of equations, a comprehensive approach in the spirit of Hamiltonian mechanics. W. Rowan Hamilton the father of post modernism, is famous for the Hamilton's variable H connected to the Lagrangian by the Legendre transform, involving momenta defined by the partial derivatives of the Lagrangian with respect to velocities. The Lagrangian defines an action through the motion of say a particle in a field, while the actual trajectory selected by nature follows the principle of least action of Maupertuis [13]. The trajectory of such particles is obtained by minimizing the quadratic variation between the kinetic and potential energies over infinitesimal time increments, leading to the classical definition of an action as the time integral of the Lagrangian.

In its bivariate representation, the Hamiltonian expresses the conjugate forms of energies of a system, where its partial derivatives with respect to the coordinates constitute Hamilton's equations, as generated by the separation of the links of the chain rule, applied to a set of coordinates of the reference frame with time t, and variable H, and is the result of a coupling between the sum of its partial derivatives and formal definition.

The Planck's constant refers to a quantity h appearing in black-body radiation linking the temperature of a source to the spectrum of its radiance, which is based on a collaborative effort between various research institutes on a proportionality between the power radiated by a source and a power of its surface temperature known as the Kirchoff law. The coefficient connecting the power radiated by a source to its surface temperature raised to the fourth power, is referred to as the Stefan-Boltzmann relation that was constructed on the basis of live quantities such as Boltzmann constant, used in Maxwellian fluids the substance of the kinetic theory of gases, as refined by Ludwig Boltzmann.

The Planck's energy, a tiny element of the corpuscular theory of light, came to life with de Broglie and Einstein's photons, a theory of light quanta inspired by the vibrating strings of a lyre of Wolfgang Amadeus. Einstein imagined an experiment back in 1905, to measure that h quantity directly, by the so-called photoelectric effect [5], which years later led to the development of a photographic imaging process on silver bromide. Moreover, Einstein is known for the equivalence principle as an introduction to general relativity for the description of phenomena such as the deflection of light, where photons follow the geodesics of curved spacetime, and his work on the field equations viewed as a mean to challenge Newton's law.

The law of attraction itself is commonly related to the notion of force, strongly tied with the inertia principle in line with Newtonian dynamics, as discussed below. Moreover, the law of attraction is interactive by definition and remains invariant upon interchange of the counterparties in two-body systems, as in the reciprocity theorem. Lastly, interactive attraction in matrices may involve semi-elastic bindings and be represented as tiny vibrating elements within a substance, or be part of a lattice on a two-by-two relationship between the vertices of a n-body system, and may involve oscillatory magnets carrying spins as a support of Shannon lyras.

The Niels Bohr model is a representation of an electron revolving around an atom's nucleus, involving a h-quantity defined as the product of the electron momentum by one circumference of the ring [3, 4]. This is the model view in use to analyse the radiance of a substance through the Fraunhofer lines, representing changes in energy level of atomions by absorption or emission of photons. The emission lines are thin light rays within a spectrum representing energy transitions in classical electron rings, while absorption lines as a complementary of the former are shadow rays, involving light passing through a translucent ionic phase. As such, Lyman rays in the UV range are in use to analyse the spectrum of an ionized atmosphere forming a rainbow by a Compton refraction.

In the remaining of the manuscript, Niels Bohr model and Larmor formula as building components of the classical theory of the electron are brought together into a self-contained representation of the atom. Moreover, the connection between the gravitational constant at Earth's surface and conformal electrodynamics is developed further down. As soon as the bridge between Aether and the gravitational vault at about 6'000 feet above sea level was uncovered at Mount Wilson [7, 14], Eather drift revealed Poisson law of gravitation. Soon after the floods, Paul Dirac took off the ground and flew overseas.

2 Theoretical development

2.1 Principles behind the Niels Bohr model

Niels Bohr model is a flat representation of an atom where electrons revolve in a disk around a nucleus. There is a strong similitude between Niels Bohr's model and the Copernician model of the solar system, suggesting that the gravitational field can be modelled as an undulatory wave. The planar model of the atom assumes a radial field emanating from a nucleus as per Coulomb's law, which is viewed as an extension of Newton's law to electrical charges. The undulatory nature of the radial field induces a magnetic field of cardioidal shape, maximized in the tangential direction to the field line, see (17). The tangential component to the magnetic field causes the electron to rotate and gain kinetic energy.

In principle, the curvature of electron orbitals is responsible for the radial component of acceleration acting in opposite direction to the force of attraction between the electron rings and respective nuclei. The question arises, why would nuclei stay in the center? As per Galileo, the inertia principle, which is the resistance of a physical object to any change in its velocity vector, causes the most massive element of a n-body system to be at the center. Inertia is also the principle causing the curvature of the trajectory of say a satellite in orbit around a massive object to induce the radial component of acceleration, preventing it from falling vertically.

To be considerate of the hydrogen atom, the mass ratio between the nucleus and its electron counterpart is a critical element of electrodynamics and quantum theory. In essence, if the mass of a nucleus is infinitely larger than its revolving electron, then we can say the nucleus is a central vertex. For two-body systems, the decrease in the mass ratio between a nucleus and perihelion, produces an eccentricity causing the nucleus to rotate inside the atom forming a tiny ovoid.

When applied to atoms, the eccentricity of nuclei may produce slight anomalies of electron orbitals. In the Bohr's model, the angular momentum of an electron is an integer multiple of the reduced Planck's constant, and electron speed is determined by matching the centripetal force to Coulomb's law. The ground floor of the hydrogen atom occurs when the integer multiple associated with the reduced Planck's constant is equal to one. Energy levels above ground state occur by absorption of photons until reaching the apex at the ionization limit of an atom. The transitions between energy levels of the Bohr model are in use in Rydberg formula for the study of the wavelengths of line spectra of atoms, where the primordial element is defined as single nucleus bound to an electron.

2.2 The quadratic wave function and connection with Coulomb's law

Given a wave function $E(r,t) = E_s \cos(w\,t)$ moving in a radial direction where E_s is the amplitude and w the wave frequency, the energy flux is equal to the time derivative of the squared wave function. This is the quadratic wave function, which was introduced to account for the 3-D dimensionality of space as a propagation medium for the wave. Hence, we get:

$$S = \frac{\partial}{\partial t} E^2(r,t) = w\, E_s^2\, 2\sin(wt), \qquad (1)$$

which is the equation of the electron wave-particle duality. For electromagnetic waves, the frequency is expressed as $w = c/\lambda$ where c is the speed of light and λ the wavelength. The effective value by the root-mean square, leads to the relation:

$$S = w\, E_s^2. \qquad (2)$$

As the flux S is expressed in joules per second and square meters and the wave frequency w is an inverse of time, the amplitude of the quadratic wave function E_s^2 is expressed in joules per square meters and represents a surface energy density.

Energy transport within an atom may occur by conduction resulting from the gradient of the energy density, or by radiative mode by the velocity of the wave function. As the second mode is usually the most efficient, the expression for the transverse energy flux is $j = cU$ where c is the speed of light and U the energy density of radiance. Say the atom is an essential singularity, emitting energy at a given frequency from its core. Hence, the flux of energy radiated is as follows:

$$S = \frac{L}{4\pi r^2}, \qquad (3)$$

where L is the power radiated at the source and r the radial distance. This equation is commonly referred to as the inverse-squared law for energy transmitted through radiation. It provides some ground for the proportionality of an E-field with the inverse of the radius, in spherical coordinates. Although, the two modes of energy transport above mentioned are quite common, other modelling approaches involve oscillatory energy waves based on non-stationary variables as in Schrõdinger equations.

As the amplitude of the wave function E_s is independent from the frequency by definition, we set $L = w\, I$ into (3). Combining (2) with (3), we get:

$$E_s^2 = \frac{I}{4\pi r^2}, \qquad (4)$$

where the variable I represents a quantity of energy equal to the power L radiated by the source divided by the wave frequency w.

The energy flux of an electromagnetic field as given by Poynting theorem is as follows:

$$S = c\,\varepsilon_0\, E^2 = c\left(\varepsilon_0 E_0^2 + \frac{1}{\mu_0} B_0^2\right). \qquad (5)$$

From Maxwell's equations, the electromagnetic field may be split between its electric and magnetic components as per Faraday and Ampere's laws. The magnetic field B_0 is related to its electric field E_0 by the relation $B_0 = \frac{E_0}{c}$, where $c^2 = \frac{1}{\varepsilon_0 \mu_0}$ and c is the speed of light, ε_0 the permittivity of a vacuum, and μ_0 the magnetic permeability. Standard units for the magnetic and electric fields are the Tesla and Volt per meter respectively. The letter E as defined by Poynting theorem in $S = c\varepsilon_0 E^2$ is the standard wave of an E-field.

As an attempt to connect the quadratic wave function described in (1) and (2) with the standard wave of an E-field in Poynting theorem, the radial component of the wave function of the source was introduced in (3) and (4). In standard electromagnetism, energy transport by radiative mode can be represented either by a surface energy density of a cross section to the flux or by a energy density of radiance. As E_s^2 is a surface energy density perpendicular to the flux, it is linked to the energy density of radiance $U = \varepsilon_0 E^2$ as given by the Poynting theorem, by the expression $E_s^2 = \lambda U$, where λ is the wavelength of the source (not to confuse with the de Broglie wavelength of an electron). Hence, the quantity of energy contained over the surface of a sphere of radius r centered on the source is $I = \lambda U 4\pi r^2$ by relation (4), leading to $E^2 = \frac{I}{4\pi r^2 \lambda \varepsilon_0}$. Provided the quadratic term E^2 is in units of Volt per meter, the force induced by the electric field on the electron is qE^2, where q is the elementary charge. By matching the above electric force with Coulomb's law, we get an expression $I = Q\lambda$, where Q is the charge of the source and λ its wavelength.

By definition, the amplitude of a wave function is independent of its wave frequency, thus the variable I as defined in (4) should also be independent of the wavelength. This is apparently not the case in the above, preventing the wave function emanating from the nucleus of an atom to reach the theoretical Energy = Frequency × Amplitude relation. An aspect related to this anomaly, is that Poynting's theorem refers to a transverse energy flux, while the wavelength connection to the wave function is a radial component of the source. For wave functions propagating in 3-D space, the quantity of energy contained in a spherical surface centered on the source, is independent of the radius of the sphere and is indeed equal to the charge of the source, i.e. $I = Q$, as a variant of Gauss's law. The contravariant is given by the standard wave of the Poynting theorem, which can be related to Gauss's law by the unit relation $\lambda \varepsilon_0 = 1$, leading to $I = q/\varepsilon_0$.

In the 2-D classical representation referring to the transverse view of the electron orbital, we have $L = F_e c$, meaning that the force is equal to a power function divided by the speed of light. By substitution of the electromotive force $F_e = qE$ into $S = c\varepsilon_0 E^2$, by Poynting's theorem, we get $F_e^2 = \frac{S q^2}{c \varepsilon_0}$. By substitution of (3) into the latter, we have $F_e^2 = \frac{L q^2}{4 c \pi \varepsilon_0 r^2}$. As we substitute back the power function $L = F_e c$ into that expression, this leads to:

$$F_e = \frac{1}{4\pi\varepsilon_0} \frac{q^2}{r^2}, \qquad (6)$$

which is the classical Coulomb's force between two equal charges of opposite sign separated by a radial distance r. Coulomb's force, which is commonly related to Gauss's law, is the radial component of electrodynamics in two-body systems equipped with a center of gravity. By definition a force is related to the mass of an entity by the principle of inertia as per Newton's first and second law $F = ma$, and is proportional to the gradient of a potential energy induced by the source.

The aim of the below is to identify the variable, which satisfies Fourier heat conduction in spherical coordinates, expressed as follows:

$$\frac{\partial Z}{\partial t} = D \left(\frac{\partial^2 Z}{\partial r^2} + \frac{2}{r} \frac{\partial Z}{\partial r} \right), \quad (7)$$

where Z is say a state variable, r the radius, t the time and D the diffusivity.

We suppose that Z is a stationary variable, meaning that $\frac{\partial Z}{\partial t} = 0$. We set $Z = UV$ where $U = \frac{A}{r^2}$ and A is a constant. As we have $\frac{\partial Z}{\partial r} = -\frac{2A}{r^3} V + \frac{A}{r^2} \frac{\partial V}{\partial r}$ and $\frac{\partial^2 Z}{\partial r^2} = \frac{6A}{r^4} V - \frac{4A}{r^3} \frac{\partial V}{\partial r} + \frac{A}{r^2} \frac{\partial^2 V}{\partial r^2}$, by substitution into (7) we get:

$$r^2 \frac{\partial^2 V}{\partial r^2} - 2r \frac{\partial V}{\partial r} + 2V = 0. \quad (8)$$

The differential equation (8) leads to a solution of the form $V = Br + Cr^2$ where B and C are integration constants. Hence, we get:

$$Z = \frac{AB}{r} + AC. \quad (9)$$

This variable describes the radial dependence of a potential energy in spherical coordinates. As the potential energy is the radial component of the Hamiltonian in spherical coordinates, the kinetic energy is the tangential component say for a particle revolving around a nucleus. Both the potential and kinetic energies of a mass are tied together, by the so-called chain rule. We have:

$$\begin{aligned}
\frac{1}{2} m v^2 &= \int_0^t m \dot{v} \, v \, dt \\
&= \int_0^t m a \frac{\partial r}{\partial t} \, dt \\
&= \int_{r_0}^{r_t} m a \, dr,
\end{aligned} \quad (10)$$

where v is the velocity, m the mass, a the acceleration. By Newton's law, we have $F = m a$. Furthermore, as we suppose equilibrium is reached when $t = \infty$, we have:

$$\frac{1}{2} m v^2 = \int_r^\infty F \, dr. \quad (11)$$

As the force is the gradient of the potential, the law of motion at equilibrium is given by equality between the potential and kinetic energies. The force resulting from the gradient of a potential energy produces a mass to accelerate according to Newton's law $F = m a$, which is at the heart of classical mechanics, as the G-field defined in gravitation is set to match the acceleration by definition. Though an equivalent G-field can be defined for charged particles in spherical coordinates, e.g. by dividing the Coulomb's force by the mass of the particle, the standard approach is to define a Q-field denoted E such that Force $= q E$, which applies to electric fields. As shown above in (7), (8) and (9), the potential energy is a stationary variable satisfying Fourier heat conduction in spherical coordinates, a character of the law of attraction as per Newtonian dynamics.

2.3 Reconciliation of the gravitational constant with electrodynamics

This portion of the manuscript is dedicated to the reconciliation of electron representation with Newtonian gravitation, as per conformal electrodynamics. The below is based on the premise that the connection between the electric and gravitational fields is the acceleration as set by Newton's law and involves certain adjustments for consistency with respect to units and model views, and to account for the distinction between the classical and quadratic wave functions.

The conformal permittivity is defined as the theoretical permittivity obtained from Coulomb's law for an electron revolving at a radial distance equal to one Earth's radius from a nucleus of opposite charge, assuming a unit acceleration on the basis of Newton's law. The conformal permittivity as defined above is expressed as follows:

$$\varepsilon_e = \frac{q^2}{4\pi R_e^2 m_e}, \tag{12}$$

where q is the elementary charge of an electron, R_e the radius of the Earth and m_e the mass of the electron.

Say the gravitational acceleration at Earth's surface is expressed as follows:

$$g. = \frac{q}{\sqrt{m_e\,\varepsilon_e \varepsilon_0\,N_A}}, \tag{13}$$

where q is the charge of the electron, m_e the mass of the electron, ε_e the conformal permittivity as defined above, ε_0 the standard vacuum permittivity, and N_A the Avogadro's constant. Avogadro's constant is in use in (13) for the scaling of the permittivity commonly expressed in Faraday per meter, where the Faraday represents a unit of charge in a mole of electron.

Combining (12) and (13), we obtain:

$$g. = R_e \sqrt{\frac{4\pi}{\varepsilon_0\,N_A}}, \tag{14}$$

which is the conformal gravitational constant, where R_e the radius of the Earth, ε_0 the vacuum permittivity, and N_A Avogadro's constant. Eq. (14) is also related to the field equation $\phi_E = \frac{q\dot{v}^2}{4\pi r^2}$, which is a primitive form having a similitude with Gauss's theorem.

2.4 Larmor formula

The Larmor formula is used to compute the power radiated by a non-relativistic charged particle as a result of acceleration. A formula for the loss of energy radiated to accelerate a charged particle from rest up to a given velocity is provided in the original work of Larmor [12]. The classical Larmor formula is disclosed in more recent works such as for the Bremsstrahlung effect and the study of electromagnetic radiations emitted in cyclotrons, a type of particle accelerators.

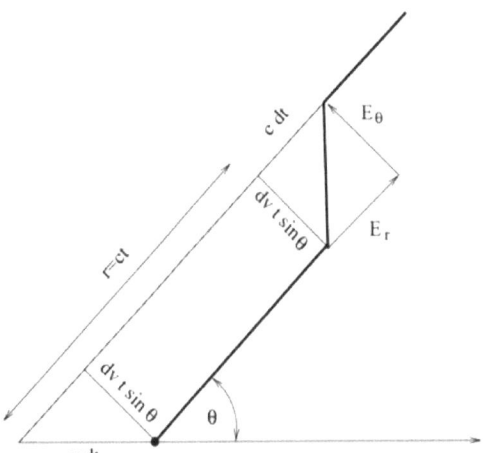

Figure 1: E-field in the region of an electromagnetic pulse in polar coordinates.

The electromagnetic wave as a bimodal function is often represented as two undulatory waves moving in the same direction, which functions are orthogonal by the inner product. The magnetic cardioid is a geometric representation, involving the interaction between an electron and an electric field. This is one of the bases for the wave representation of the electron to be so useful. The letter B, which commonly refers to a magnetic field, is represented by an E-field in the current context, denoted by E_θ, where θ is the angle between the radial electric field E_r and the orientation of E_θ itself. As the magnetic field is maximized in the tangential direction to the motion of the electron, we can infer that E_θ is a $\sin\theta$ multiple of E_r.

From the quadratic wave model in (2), we have $E_s^2 = \frac{S}{w}$, where S is the energy flux and w the wave frequency. As we suppose that the E-field is proportional to the inverse of the wave frequency, where the ratio of wave frequencies is equal to the ratio of velocities, we get $\frac{E_\theta}{E_r} = \frac{v_r}{v_\theta}$. By Pythagorean theorem, we have:

$$\frac{E_\theta}{E_r} = \frac{\Delta v t \sin\theta}{c \Delta t}, \qquad (15)$$

where E_θ and E_r are respectively the tangential and radial components of the E-field, c the speed of light, and t the time after a pulse Δt. From equality $\frac{v}{t} = \frac{dv}{dt}$, we get $v\Delta t = t\Delta v$. By definition, t is the time to accelerate a charged particle q from rest to velocity v.

The radial component of an E-field as in Coulomb's law, can be expressed as follows:

$$E_r = \frac{q}{4\pi\varepsilon_0} \frac{1}{r^2}. \tag{16}$$

Given the acceleration term $\dot{v} = \frac{\Delta v}{\Delta t}$ and joint relation $r = ct$, (15) and (16) lead to:

$$E_\theta = \frac{q\,\dot{v}}{4\pi\varepsilon_0 c^2\, r} \sin(\theta). \tag{17}$$

By Poynting theorem, i.e. $S = c\,\varepsilon_0\, E^2$, we can write:

$$S = \frac{1}{16\,\pi^2\, c^3\,\varepsilon_0\, r^2} q^2\, \dot{v}^2\, \sin^2\theta. \tag{18}$$

The angular element in spherical coordinates is $d\Omega = r^2 \sin\theta\, d\theta\, d\varphi$, leading to the below expression for the power radiated by an electron:

$$P = \int_{\theta=0}^{\pi} \int_{\varphi=0}^{2\pi} S\, r^2\, \sin\theta\, d\theta\, d\varphi. \tag{19}$$

As $A = \int_{\theta=0}^{\pi} \int_{\varphi=0}^{2\pi} \sin^3\theta\, d\theta\, d\varphi = \frac{8\pi}{3}$, we obtain:

$$P = \frac{8\pi}{3} \frac{q^2 \dot{v}^2}{16\pi^2 c^3 \varepsilon_0}, \tag{20}$$

which is the classical Larmor formula for the power radiated by an accelerated charge, in say Watts per squared steradians where the variables in argument are expressed in the international system of units.

When θ spans all angles in $[0, 2\pi]$, the E-field E_θ in (17) describes a cross section of the magnetic field induced by collinear E_r field lines (Fig. 1), which in 3-D can be represented by a Horn Torus. This is the doughnut representation of an electron orbital of an atom, see Fig. 2.

2.5 From Thomson cross section to the classical Planck's constant

Considering an E-field where the field lines are collinear and pulsate in the direction orthogonal to the orbital of an electron revolving around a nucleus, the energy flux over a cross section σ_e, which is the transverse power inflow, is given by:

$$P_{in} = c\,\varepsilon_0 E_r^2\, \sigma_e, \tag{21}$$

where the energy flux is the speed of light times the energy density as given by Poynting's theorem.

The power radiated by a ground state electron revolving around a nucleus is given by Larmor formula, which can be expressed as follows:

$$P_{out} = \frac{8\pi}{3} \frac{q^2\, (q\, E_r/m_e)^2}{16\,\pi^2\, c^3\,\varepsilon_0}. \tag{22}$$

As $P_{in} = P_{out}$, (21) and (22) lead to the well-known Thomson cross section for a free electron in orbital:

$$\sigma_e = \frac{8\pi}{3} \left(\frac{q^2}{4\pi\varepsilon_0 m_e c^2} \right)^2. \qquad (23)$$

By the third mass scaling rule, we multiply (23) by $(m_p/m_e)^2$, a scaling of the Thomson cross section to the Bohr sphere[i], yielding:

$$\sigma_0 = \frac{8\pi}{3} \left(\frac{q^2 m_p}{4\pi\varepsilon_0 m_e^2 c^2} \right)^2. \qquad (24)$$

Given Bohr radius $r_0 = \frac{\varepsilon_0 h^2}{4\pi^2 m_e q^2}$, the area of the Bohr sphere $4\pi r_0^2$ is expressed as follows:

$$\sigma_s = \frac{\varepsilon_0^2 h^4}{4\pi^3 m_e^2 q^4}. \qquad (25)$$

The Bohr radius r_0 represents the radius of an electron orbital, rescaled for consistency with the standard wave as defined in the Thomson cross section. The Bohr model is based on the electron identity $n\frac{h}{2\pi r} = m_e v$, where h is a constant defined as the product of electron momentum by one circumference of the ring.

By matching the Bohr sphere σ_s in (25) with the cross-sectional area σ_0 in (24), we obtain the one circumference momentum of the electron, also known as the classical Planck's constant[ii], expressed as follows:

$$h. = \frac{q^2}{c\,\varepsilon_0} \sqrt{\pi} \sqrt{\frac{2}{3} \frac{m_p}{m_e}}, \qquad (26)$$

which is based on the electron representation as per Bohr model and Larmor formula, where q is the elementary charge of the electron, m_p the mass of a proton, m_e the mass of an electron, ε_0 the vacuum permittivity, and c the speed of light.

3 Conclusion

The present manuscript aims to reconcile gravitation and light quanta with electrodynamics, a science describing the motion of electrons in a conductor as a result of

---

[i] In the membrane representation of the atom, the electron is a surface represented by the Bohr sphere, where the flux of energy radiating through the membrane is determined by the mass of the proton. The flux of energy traversing the electron surface is connecting both poles of the atom having the shape of an apple.

[ii] A precise matching with actual measurements of h is obtained with the introduction of the relativistic mass of the electron in the Bohr model, and a mass gap adjustment for the binding energy of the electron orbital

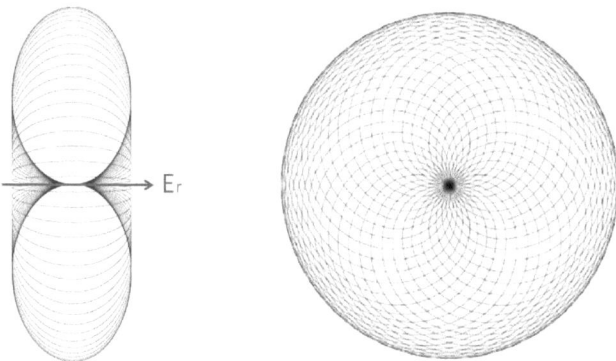

Figure 2: 2-D cross sections of an electron orbital in the doughnut representation. *Source: Horn Torus by Wolfgang W. Däeumler.*

an electric field. The classical representation of the atom in the context of conformal electrodynamics, is a modelling approach in line with certain principles related to the so-called ring theory, a flat representation of the atom based on the following axioms: (i) Each atom has a single nucleus at its core acting as a central vertex. A central vertex can be viewed as a stationary nucleus as per the inertia principle, which stems from the idea that the mass of nuclei is infinitely larger than the mass of their electron counterparts. (ii) The electrons revolving around the nuclei of atoms form rings centered on the origin of a coordinate system. These rings represent the trajectories of electrons, assumed to follow uniform circular motions. (iii) The local coordinate system in 3-D vector space in use for the motion of an electron ring, has the central vertex of the atom as its origin.

The differences between the expected and theoretical values of model variables as finite quantities, such as gravitation and Planck's constant as defined in the present work, may be attributed to deviations from uniform circular motion of electrons around the nuclei of atoms as in the second axiom of ring theory. The decrease in the mass ratio between nuclei and electron counterparts in two-body systems, may induce the rotation of atoms' nuclei in small ovoids, which eccentricities may cause electron departure from uniform circular motion. A more accurate representation of ring theory from a metrological viewpoint, would incorporate such eccentricity in use for calculations connext to the perihelion and aphelion of planetary motions as in the classical representation of the precession of Mercury, which obliquity can be viewed a an oval having an interest in general relativity and former observations [8].

As an example of the views of the electron presented in various sources, the Larmor formula is a 2-D representation of the magnetic field resulting from the adjacent cross section of an electron orbital, while Bohr model represents a transverse cross section of the forming atom. The combination of the two views of the atom into a single equation leads, to the classical expression of the Planck's constant.

The latter is obtained for an elementary corpuscle composed of one proton and a single electron in its ground state, and by adding flavors can be extended to atoms with many electrons or nucleons. Moreover, the gravitational attraction arises from a primitive form of field equation related to Gauss's law as applied to an elementary charge by the molar fraction, and is connected to conformal permittivity as defined for an electron revolving at one Earth radius from its nucleus on the basis of a unit acceleration of the ring as given by the axion in the second Newton's law.

4 References

[1] AMPÈRE, A.-M.: *Description d'un appareil électro-dynamique construit par Ampère*, Les librairies Crochard et Bachelier, 1824

[2] AMPÈRE, A.-M.: *Théorie Mathématique des Phénomènes électro-dynamiques uniquement déduite de l'expérience*, Hermann libraries 2nd ed, drafted in 1826, 1883

[3] BOHR, N.: I. On the constitution of atoms and molecules. In: *Philosophical Magazine, Series 6* 26 (1913), S. 1–25

[4] BOHR, N.: XLII. On the quantum theory of radiation and the structure of the atom. In: *Philosophical Magazine, Series 6* 30 (1915), S. 394–415

[5] BROGLIE, L. de: XXXV. A tentative theory of light quanta. In: *Philosophical Magazine, Serie 6* 47 (1924), S. 446–458

[6] COULOMB, C. A.: *Premier, second et troisième mémoire sur l'électricité et le magnétisme*. S. 569–638, Histoire de l'académie Royale des Sciences, 1785

[7] DIRAC, P. A. M.: Is there an Aether ? In: *Nature* 168 (1951), S. 906–907

[8] EINSTEIN, A.: Explanation of the Perihelion Motion of Mercury from General Theory of Relativity. In: *Königlich Preußische Akademie der Wissenschaften, Collected papers, Princeton*. (1915), S. 831–839

[9] FARADAY, M.: *Experiental researches in Electricty*, Philosophical transactions, 1839

[10] FARADAY, M.: *Experimental Researches in Chemistry and Physics*, Taylor & Francis, 1859

[11] HIGGS, P. W.: Broken Symmetries and the Masses of Gauge Bosons. In: *Physical Review Letters* 13 (1964), S. 508–509

[12] LARMOR, J.: LXIII. On the theory of the magnetic influence on spectra; and on the radiation from a moving ion. In: *Philosophical Magazine, Serie 5* 44 (1897), S. 503–512

[13] MAUPERTUIS, P. L. M.: *Les Loix du mouvement et du repos déduites d'un principe métaphysique*. S. 267–294, Académie Royale des Sciences et des Lettres, 1746

[14] MILLER, D. C.: Ether-Drift Experiments at Mount Wilson. In: *Proc. Nat. Ac. Sci., USA* 11 (1925), S. 306–314

[15] ØRSTED, H. C.: New experiments by Dr. Seebeck on electromagnetic effects. In: *Annales de Chimie et de Physique* 22 (1823), S. 199–201

[16] REINES, F. ; COWAN, C. L.: Neutrino physics. In: *Physics Today* 10 (1957), S. 12–18

[17] RUTHERFORD, E.: The scattering of α and β particles by matter and the structure of the atom. In: *Philosophical Magazine, Serie 6* 21 (1911), S. 669–688

[18] THOMSON, J. J.: XL. Cathode rays. In: *Philosophical Magazine, Series 5* 44 (1897), S. 293–316

[19] THOMSON, J. J. ; RUTHERFORD, E.: XL. On the passage of electricity through gas exposed to Röntgen rays. In: *Philosophical Magazine, Serie 5* 42 (1896), S. 392–407

Connection between the cosmic horizon and the ladder to the hypersphere

Yuri Heymann, winter 2020.

Abstract The Parthenon sequence is an abstract representation of space geometry composed of elementary shapes of the ring as a continuum of shape functions. As a reverse holographic effect of Moon's shadowing, the ladder to the hypersphere can be viewed as a metamorphosis leading to an expansion of the 3-D sphere, by some equivalence of a quantity, as a kind of homeomorphism defined in Poincaré conjecture. We further provide the ground for the light-square relation, a fundamental relation in cosmology linking Minkowski radius to the cosmic horizon, by a light beam of cylindrical shape, leading to Hubble's constant expressed as a function of Bohr radius and photon sphere.

1 Introduction

This was a long time ago, back in the early times, when Euclid traced a straight line on a tablet of clay, giving birth to the first character of the alphabet. Then with a leaf, added a second line forming a syllable, both adjacent to each other, and having both ends touching each other at infinitum. As sharp as a fork of zircon anchored in the limes, with his arm Pythagoras drew a right triangle, having an arrow at the opposite side pointing towards the zenith, to project a point in the direction of Venus to the floor. In the meanwhile, Rayan placed a straw made of chalk on Euclid tablet, and by summing up the interior angles on either sides of the intersects with both Euclidean lines, opened his mouth spanning a hemisphere of 180°, equal to half a quarter of minutes of the Sundial. A few years later, Arthuro of Carthage drew two concentric circles, representing spheres of arbitrary radius on a 2-D arity of the Euclidean space, such that the lengths of any two handles pointing at Polaris in the sky, are normalised by a modulus obtained, by orthonormal projection of the line of sight, onto the base of a right triangle. As a ratio, the lengths of both handles pointing at the sky, divided by the lengths of their shadows at the zenith, are equal by a proportion giving rise to trigonometry. Moreover, this geometry as provided with a norm and a right angle can be extended to non-Euclidean geometry, in line with the Pythagorean school of thought. To materialise the axioms of Euclid, back in Thessaloniki, Aristotle drew five concentric spheres, the celestial sphere in first instance, the zodiacal constellations, the heaven's vault, the planets of the solar system and Earth, and the stars of the Milky Way.

This was the beginning of a new era in Hellenica, when Eratosthenes the geographer developed a system to measure the distance between nearby cities, on the basis of a relation between the travel time of a secant and angular diameter, as a reference to cartography in Ptolemy time, giving rise to a stere as a subdivision of the circle referring to the separation between Athens and Alexandria, a standard to build new axes on curved surfaces. As a prerequisite of the foregoing, the relation among size and distance of the Sun, Moon and Earth, were based on various observations involving solar and lunar eclipses, sharing a similitude to parallax in use to measure distances in the solar system. The prevailing view, supported by Eratosthenes, was that the perfect sphere consisting of a contour containing a certain volume, is characterised by a center and a measure of radial distances. From Carthage to Aristotle, the surface area of the sphere defined as the full set of equidistant points separating its contour from its center of gravity became the standard. Pluton separated the ores from the rare, giving birth to a new planet, the last planet of the solar system.

A relevant aspect observed during Ptolemy period, is the equant problem related to an observational skew, between the distance measurement separating two cities using an equal subdivision of the Sundial, and the parallax for the calculation of stellar distances. The relation between events separated in spacetime, is a special domain in relativity, which led to a further refinement of the Sundial in Constantine time, by giving an obliquity to the gnomon of an oval day, to match an equal subdivision of the quadrant during the epagomenal days. This device was in use for measurements involving various effects such as seasonal variations and obliquity of Earth.

As Aristotle was back in the Acropolis from a journey in Macedonia, he had his guests gathered around the banquet and handed some bred to Sunday. Sitting on his side, Lucy one of his fellows defined a shape as something that can contain a liquor, and Demeter as something that can be harvested. The Parthenon sequence referred in the remainder of the manuscript is an abstract representation of space geometry composed of a succession of shapes derived from the sphere. Arturo of Cartage defined some kind of algebraic functions, as an expression to describe a shape or be derived from another shape by say a morphism, or be part of a whole. As such the Parthenon sequence is made of elementary shapes forming a set of antithetic varieties, say a balloon containing a certain quantity of points as in ideal fluids, not to say a surface enclosing a stere of molecules such as oxygen, or to be an open surface, the contour of a surface, an empty set and vice versa. The continuum of Parthenon sequence of space geometry beyond Euclidean geometry, is an opening giving rise to new forms and potentials, on the way to the heaven's vault from the Himalayan to Eathers. In a quest for harmony between the sky and Earth, Icarus went down the Hills and discovered love in a leaf. From the openness of the limbs, Neptune submerged out of the scum, giving rise to salty water, flowing from the Aegeans to the fresh waters.

In molecular representation of aquiferous solutes [3, 5], as in diatomic representation of fresh water, "La rotondité" of a perfect sphere, is asymmetric, in configurations when vertices connecting its nuclei are not evenly distributed around its surface, whereas symmetry is critical in holographic expansion of hypersphere, not to say in configurations where vertices are convergent as in inertia of rotating bodies. In the above representation, asymmetry is the principle leading to many configurations as in the Heavenly spheres of geometry. The gyroscope is a kind of sphere having many rotational axes embedded within a vault made of concentric rings. As such a biaxial gyroscope, is an example of sphere having four degrees of freedom, a peculiarity of 4-D space geometry, as a character of electron spins and a principle behind flavors of molecular representation of fluids such as hexane and other volatile substances.

The azimuth as angle between the North pole and a point on the surface of a sphere, is in use to project the ephemerides on the x-y plane perpendicular to Polaris axis. The standard approach to the 4-D world view, consists of matching the degrees of freedom of a 3-D coordinate system, say in spherical coordinates (O, r, θ, φ), where the origin of the frame O is moving in linear fashion over time t, as in inertial reference frames. In ring theory, the Carrousel representation, is another candidate based on classical sphere as a reference frame $(O_2, r, \theta, \varphi)$, where O_2 is the origin in a circular rotation around another center O_1, where the position of O_2 is described in polar coordinates in a primary frame (r_1, ω). This is a key ingredient in classical mechanics related to Coriolis force, in use among other things to describe planetary motion of electron rings and as an element in Ptolemy system. Then came the thrust inspired by Ptolemy model, as a design connecting a power supply to a disk in rotation by a thin rod. This device is in use in engines for propulsion say on manifolds as in parallel transport, e.g. steam locomotive force [4]. As a special case, the Minkowski spacetime is referring to the 4-D universe, where events are inscribed in two opposite light cones, which intersection indicates present time.

2 The Parthenon sequence of space geometry and shape compositions

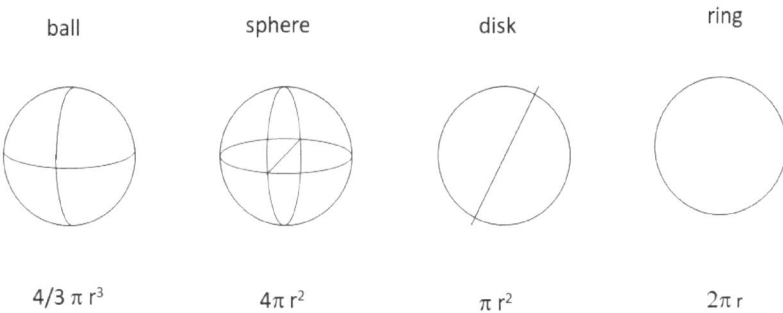

Figure 1: The four elementary shapes of the ring in the sequence of space geometry

The Parthenon sequence is an abstract representation of space geometry composed of elementary shapes of the ring as a continuum of shape functions. Say ν stands for the degrees of freedom of space, and χ for the degrees of freedom of geometry. A third axis orthonormal to the ground floor denoted ζ, stands for the height of geometry, which by incremental moves indicates stairs in the Parthenon. The consecutive sequence of four elementary shapes (Fig. 1) are ranked by some dimensions as attributes of the shapes, which in the current context represent a perimeter, surface or volume. Moreover, the surface of the sphere is obtained by applying the partial derivative with respect to radius to the volume of a ball, and the circumference of the ring by the same operation to the surface of a disk.

When reading the shapes from right to left, a discontinuity appears between the second and third shapes, not following the chaining by the $\partial/\partial r$ operator as applied to shape functions. By the quotient rule, the disk and the sphere are joined together with a surface ratio linking both shapes. This connection is a kind of horizontal continuity by an action of say cosmic rays, which enables the prolongation of the Parthenon to rank zero, by such projection on a one-dimensional vertical line adjacent to the ring. By shape composition, the inner product between any two shapes of the sequence as defined below, leads to a new shape of that sequence. Say the inner product between any shapes of the sequence is defined as follows:

$$\langle f, g \rangle = \int_0^1 \psi \, f(r) \, g(r) \, dr, \qquad (1)$$

where $f(r)$ and $g(r)$ are consecutive shape functions, ψ a potential and r a radial coordinate. The integration between 0 and 1 stands for normalisation, where the module of a squared-integrable function $f : \mathbb{R} \to \mathbb{R}$ on the real domain is defined as $||f|| = \sqrt{\langle f, f \rangle}$. Say the potential associated to the product rule of consecutive shapes is expressed as $\psi = \frac{1}{\phi r}$, where ϕ is a real variable independent of radius. On the side, the chaining rule by the $\partial/\partial r$ operator, stems from a specific potential given by $\psi = \frac{1}{4\pi r^2}$, where r is the radius.

A special case arises when two consecutive shapes share the same exponent as in the disk-sphere connection, as given by $f(r) = \pi r^2$ and $g(r) = 4\pi r^2$ and where n is the exponent of say the first shape by definition. Given the surface ratio $\&^2 = 4$ between the disk and the sphere, by an extension of the potential of the chaining rule, we get $\psi = 4 \times \frac{1}{4\pi r^2}$, leading to $\langle f, g \rangle = \frac{4\pi}{3}$ as the 4th unit of a ball. Say $\phi = 2\pi$, i.e. $\psi = \frac{1}{2\pi r}$, giving $\langle f, g \rangle = \frac{\pi}{2}$ a quarter of a pie, which was the first step of Euclid beyond the Parthenon. Uh, Hay said $\&^4 = 16 \times \frac{\pi}{2}$, as $8\pi r^4$ of a bracket.

Shay said a sharp function representing a n-sphere of radius r, expressed as $V_n = v_n r^n$, where v_n stands as a shape unit, and n an index. As Cartès, a hypercube of unit shape factor, which by a 2π aperture be represented as a 4π solid angle, as per say.

As a closed shape containing a quantity of a substance, the surface of a 3-sphere is homeomorphic to a Riemann sum over infinitesimal elements as a measure of distance separating events, which surface represents a likelihood by some smooth function. As a set of tiny elements touching each other in a continuum, their aggregate approaches a finitude by a radius, which horizon is determined by the particle, the smallest element of that continuum.

The distribution homeomorphic to the ring, shall be composed of elements which by an inner product defines a commutative ring, of a function which derivative is equal to itself. By the transcendental numbers, we have:

$$\int_{-\infty}^{\infty} e^{-x^2} \, dx = \sqrt{\pi}. \tag{2}$$

The above function as a unit can be extended to many dimensions, through a multitude of inner products of that function, leading to:

$$\left(\int_{-\infty}^{\infty} e^{-x^2} \, dx \right)^n = \int \int \int_{\Omega} e^{-(x_1^2 + x_2^2 + \ldots + x_n^2)} dx_1 \, dx_2 \ldots dx_n. \tag{3}$$

By the extension of the 3-D world of Cartesian coordinates to many dimensions, the square radius is expressed as the quadratic sum of variables, which axes belong to a base, leading to $r^2 = x_1^2 + x_2^2 + \ldots + x_n^2$, where x_i for $i = 1, \ldots, n$ and $n \in \mathbb{N}$. The connection between Cartesian coordinates, and say an element of Sphere, is given by $dx_1 \ldots dx_n = r^{n-1} \, dr \, d\Omega_{n-1}$, leading to:

$$\left(\int_{-\infty}^{\infty} e^{-x^2} \, dx \right)^n = \int \int \int_{\Omega} e^{-r^2} r^{n-1} \, dr \, d\Omega_{n-1}. \tag{4}$$

The unit element of say a n-sphere is defined as $v_n = \int \int \int_{\Omega} d\Omega_{n-1}$, where $d\Omega_{n-1}$ is a tiny element of that geometry. By variable interchange say $u = r^2$ in the above, leads to:

$$v_n = \frac{\pi^{\frac{n}{2}}}{\Gamma\left(\frac{n}{2} + 1\right)}, \tag{5}$$

where Γ is the Gamma function defined as $\Gamma(z) = \int_0^{\infty} x^{z-1} e^{-x} \, dx$ where $\Re(z) > 0$ and $n \in \mathbb{N}$. Eq. (5) is the standard n-sphere, which by inner product leads to the quadratic radius resulting from square-rule composition, forming a frame for the prolongation of the elements of the diagonal of the Parthenon (Fig. 2).

The squared radius of a sphere as the sum of three little squares of sides x_1, x_2 and x_3 i.e. $r^2 = x_1^2 + x_2^2 + x_3^2$ (see Platon's sphere), be employed as a distance measurement. In 3-D world, the quadratic expression of the radius of a sphere be equal to three faces of a cube, in a direction maximizing the norm of such vector, where the weighted sum of respective basis elements x_1, x_2 and x_3 form the edges of the cube. The quadratic radius led by square-rule composition in 3-D geometry, is a kind of cosmic projection by the cubic form of a 4-D sphere.

As such the proportionality between two radii of the Moon by the 3/2 rule (referring to 3 planisfers), i.e. $r_3^2 \sim r_4^3$ where r_3 and r_4 are radii in the 3 and 4-D views, yields an expression for the radius of a 4-D sphere as a function of coordinates raised to the third of a power. Moreover, in 3-D you can only see one hemisphere of the Moon, whereas in 4-D a third of its surface (as per the 3/2 ratio). Say in Ricci flow interconnecting a 3-D space to higher dimensions as in the curved geodesics of light rays, the 3 and 4-D views of a hypersphere laying in the cosmic fluid, can be viewed as a continuum connecting the ionosphere to the Aegeans, by the iridescence of a sight from Earth. The path taken by the lines of sight of an aperture, forming an opening of $\frac{2\pi}{n}$ radians, where $1/n$ represents a fraction of Mund sphere, spanning a

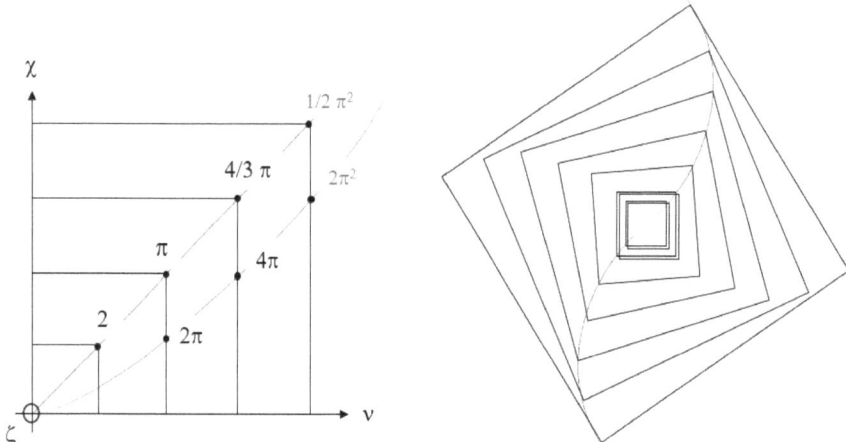

Figure 2: The Parthenon sequence of space geometry up to $\nu = 3$. The extension of the Parthenon to higher dimensionalities is achieved here by square-rule composition, on the basis of a Gaussian distribution.

light cone touching Moon's surface. At the fifth epagomenal day during the Hijri lunar day, in light of Rayan touching sea surface, the Ionian Sea and hypersphere intersect. Say R is the radius of Moon, and L its shadow as viewed from Earth, by the cross section of a Minkowski light cone at the intersect of sight with Moon's surface. By a fraction of say $360°$ as an opening of space geometry connecting the surface of a 4-D sphere to the 3-D world, we have $4\pi R^2 = n \times \pi L^2$, expressing a variety of Pythagorean theorem given by $L/R = \sin\left(\frac{2\pi}{n}\right)$. The merging of the above by the link of a joelry leads to $\sin\left(\frac{2\pi}{n}\right) = \frac{2}{\sqrt{n}}$, where $n \in \mathbb{N}$. As a leaf of a functional, this equation has a natural solution $n = 3$, belonging to the one-third equant of the Moon, the link to the 4-D hypersphere by some $L \cdot \forall/\pi$ relation.

As a reverse holographic effect of Moon's shadowing, the ladder to the hypersphere may be viewed as a metamorphosis leading to an expansion of the 3-D sphere, by some equivalence of a quantity, as a kind of homeomorphism seen in Poincaré conjecture. The 3-sphere defined in Poincaré, is referring to a hypersphere which contour is a three dimensional shape forming the boundary of a ball in 4-D. A 3-manifold on its own is a 3-D shape, which is locally homeomerphic to 3-D Euclidean space, meaning that every point has a neighbourhood approaching an open 3-D ball at an infinitesimal scale.

Although the extension of the squared norm to orthogonal set of functions as Chebychev and Legendre polynomials is enticing, it is worthwhile providing some background below.

Consider a set of two functions say $\mathcal{A} = \{a\cos t, b\sin t\}$, where a and b are scalars

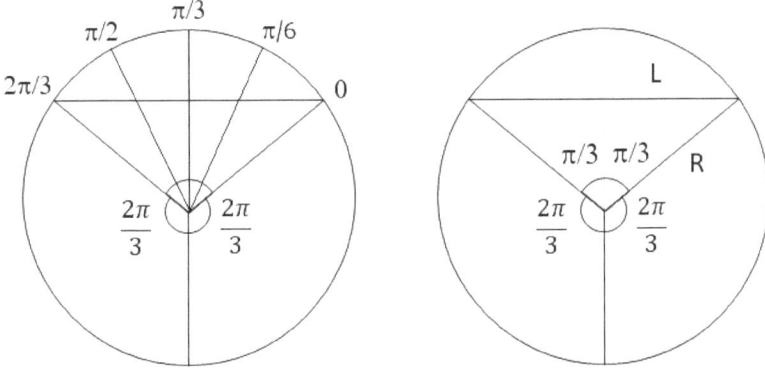

Figure 3: The one-third equant of the Moon. - *Source: The Planisfers of the Silk Atlas, 2020.*

($a > b$) and t a variable in $[0, 2\pi]$, forming an orthogonal basis of functions by an inner product, i.e. $\langle f, g \rangle = \int_0^T w\, f(x)\, g(x)\, dx$ where w is a weighting function.

Say the objective function $f \colon \mathbb{R} \to \mathbb{R}$ is expressed as the sum of the two functions of $\mathcal{A}$, leading to $f(t) = a \cos t + b \sin t$. The tuple formed of the two functions spans an ellipse for $t \in [0, 2\pi]$, where a and b are the lengths of the semi-major and semi-minor axes. The function $f(t)$ is maximized when $t = \arctan(b/a)$, leading to $\max[f(t)] = \sqrt{a^2 + b^2}$ as a brick of the squared norm in 2-D geometry. The trigonometric functions, sine and cosine are even functions, in the sense that the derivative of either one of the two is equal to the remaining, in absolute terms. Moreover, the sequence formed of successive derivatives of the sine is looping by the quarterly subdivisions of a circle.

The extension of the above 2-D norm to higher dimensions, can be envision by some countable subdivisions of the circle, introducing some trigonometric functions as elements of an orthogonal basis set, e.g. in 3-D world, say three little words such that the resulting two degrees of freedom bundle is *Lebesgue*-integrable by some inner product. The connection between the third dimension and a 2-D arity of space, may be performed by some equivalence of the cross product on the basis of existing functions, as in orthonormal vector space. In a ternary trigonometry, a common approach would be to consider a 3-D shape spanned by some weighted function of the sine and cosine in 3-D Euclidean space, e.g. $\{a\cos(t), b\sin(t), c\cos(\beta s)\}$, matching the norm expressed as $||v|| = \sqrt{a^2 + b^2 + c^2}$ by a shadowing, and where the union of the marginals is homeomorphic to a circle as viewed from a point of sight in the direction of vector v, by a stereoscopic projection.

In a normed vector space, a vector is the result of the weighted sum of its basis elements, which norm is maximized in a direction determined by some weighted distance to its sub-elements, leading to:

$$||f||^2 = \max. \left| \sum_{n=1}^{N} a_n \, q_n \, f_n(\mathcal{E}_{N-1}) \right|. \tag{6}$$

where a_n are the respective weights associated with each basis elements, f_n the orthogonal functions of each basis set, $\mathcal{E}_{N-1} = \{\varepsilon_1, \varepsilon_2, ..., \varepsilon_{n-1}\}$ a set of $n-1$ variables representing the degrees of freedom of the underlying geometry, and q_n represents a fraction of say $\{n+1\}$-manifold as visible from a n-dimensional space, in a stereoscopic projection or by some shadowing function.

By applying say a function f_n bijective on a subelement $[a, b]$ of $\mathbb{R}$, where the variabe $\tilde{x}_n$ results from a transform of $x \in [a, b]$ having the property to maintain angles unchanged, and where $\tilde{x}_n$ spans $[0, 1]$, leads to:

$$f_n(x) = \phi^{-1}(\tilde{x}_n), \tag{7}$$

the link between the marginals of a n-dimensional smooth manifold by the $\tilde{x}_n$ variables of a unit length. The character ϕ^{-1} denotes the inverse cumulative function of a standard density sharing a transcendental with the sphere. As a cumulative function of density, ϕ is the link to the statistical world of random variables and distributions.

3 Cosmic horizon in geometry of the celestial sphere

As from the alignment of Sirius with the Sphinx at the temple of Amun-Re, Thales of Milettes measured the height of pyramids in Cairo, from their shadows in the sand at the glowing of the Moon (Fig. 4).

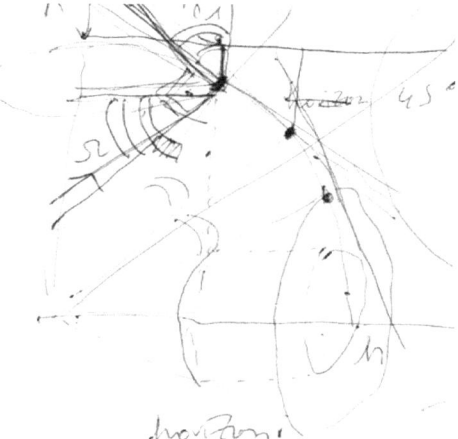

Figure 4: The aperture of Minkowski light cone and horizon - *Souce: 1976's Skylab experiment, da Vinci gallery, 2020.*

Considering a sphere of radius R, the distance to the horizon R_a from the little height h above sea level as obtained by Pythagorean theorem (Fig. 5), leads to the distance to the horizon expressed as:

$$R_a = \sqrt{2\,R\,h + h^2}\,. \tag{8}$$

When $h \ll R$, we have $R_a \sim \sqrt{2\,R\,h}$, which is the distance to the horizon from the little height h. Say $h = \alpha\,R$, then by the portion of a whole, we have:

$$R_a^2 = 2\,R^2\,(\alpha + \alpha^2)\,. \tag{9}$$

Consider a brick of height α and lenght $1 + \alpha$, fitting into a disk of unit radius. Thus, we have $\pi = \alpha + \alpha^2$. In light of the above, the relation between the horizon and cosmic radius is expressed as follows:

$$R_a^2 = 2\,\pi\,R^2\,. \tag{10}$$

Laying in the cosmic fluid made of a myriad of tiny stars in the Milky Way, the celestial sphere is suspended in Eathers (see Fig. 6). The bath of cosmic rays exerts an action on the sphere, producing a force. Say θ is the azimuth of a cosmic ray, which is the angle between the North pole and the direction of the cosmic ray. Given a light ray as viewed from Earth, the ratio of photons' energies between the source and observer, expressed as a scale factor is a function of the solid angles of a bundle connecting the source to the observer, as per the reciprocity theorem. As a quotient of surfaces defined by the action of unidirectional light rays on a sphere, the ratio of effective to geometric cross section is as follows $\Omega_s/\Omega_g = \frac{1}{2\pi} \int_0^{2\pi} \int_0^{\pi/2} \cos\theta\,\sin\theta\,d\theta\,d\varphi$, with surface element $dS = r^2\,\sin\theta\,d\theta\,d\varphi$ defined on a hemisphere. The resulting

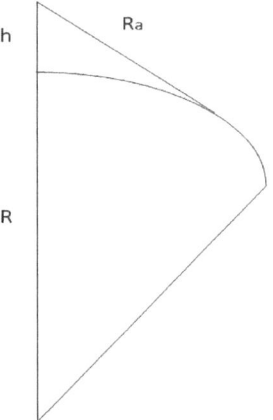

Figure 5: Horizon as viewed from the little height h above sea level

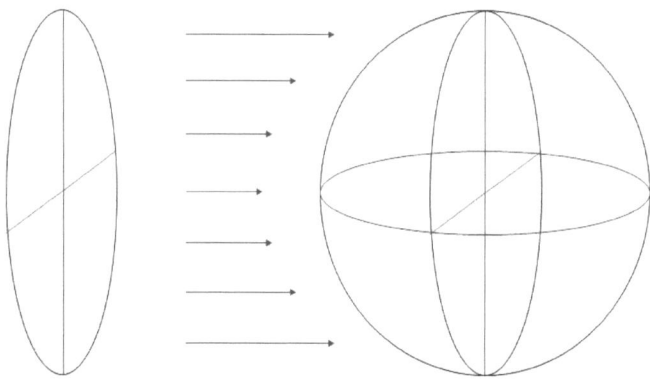

Figure 6: Cosmic rays on the left hemisphere of the celestial sphere

geometric cross section is $\emptyset = 2\pi R^2 \times (\Omega_g/\Omega_s)$, leading to the relation between the horizon of a light ray and geometric radius of the celestial sphere, expressed as:

$$R_a^2 = 4\pi R_g^2. \tag{11}$$

As the cosmic radius of a light ray, is expressed as $r_h = ct$, the light-square relation becomes:

$$(ct)^2 = 4\pi R_g^2, \tag{12}$$

where c is the speed of light and R_g the geometric radius of the celestial sphere.
Eq. (12) is a fundamental relation in cosmology connecting the Minkowski radius $r_h = ct$ to the surface of the celestial sphere. Say we have a cube and light cones with null events centered on the origin of the cube, such that each light cone intersects a face of the cube. For such light cones inscribed in a sphere of radius R_g, leads to six distinct disks of radius R_0 (Fig. 7). By matching the area of the six disks with three quarters of a light square, we get $6\pi R_0^2 = \frac{3}{4}(ct)^2$, where R_0 is the radius. As $R_g = \sqrt{2} R_0$, we get $(ct)^2 = 4\pi R_g^2$, where R_g is the geometric radius of the celestial sphere, see eq. (12).

As viewed from an atom, the cosmic sphere is spanned by a light beam of cylindrical shape, having a quadrilateral contour defined by its radius r_0 and length R_h. The area under this contour is equal to $A_0 = r_0 R_h$, where r_0 is the Bohr radius and R_h the cosmic horizon. By the mean square rule we have $R_g = \sqrt{r_0 R_h}$, which is the geometric mean between the Bohr and cosmic radii. Thus, we have:

$$(ct)^2 = 4\pi r_0 R_h \tag{13}$$

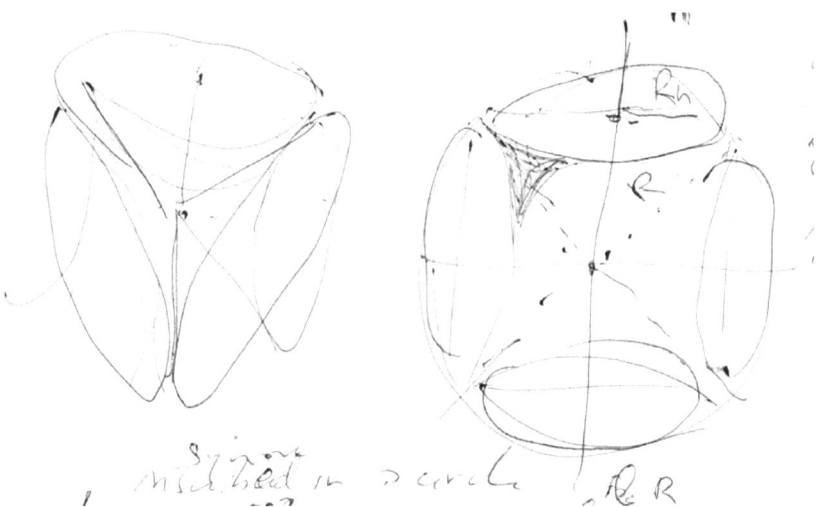

Figure 7: Cubic representation based on Minkowski light cones, radii R_0, R - Source: The Mund, on Mars landing, da Vinci gallery, 2020.

Eq. (13) is an equality between the transverse cross section of the Bohr cylinder and a quarter of light square (Fig. 8). The transverse cross section of the Bohr cylinder is defined as $\emptyset_\circ = \pi R_g^2$, where $R_g^2 = r_0 R_h$, whereas the light square as $R_a^2 = (ct)^2$. The above leads to the cosmic horizon expressed as:

$$R_h = \frac{(ct)^2}{4\pi r_0}, \tag{14}$$

where r_0 is the Bohr radius as per Niels Bohr model of the atom where an electron revolves in a disk, which by the product of the electron momentum by one circumference of the ring gives $n \frac{h}{2\pi r} = m_e v$, see [1, 2]. As such Bohr radius is expressed as $r_0 = \frac{\varepsilon_0 h^2}{\pi m_e q^2}$, where q is the elementary charge of an electron, m_e the mass of an electron, ε_0 the vacuum permittivity, and h Planck's constant.

By the relation $R_h = \frac{c}{H_0}$ where c is the speed of light of a unit time pulse i.e. $t = \delta t$ equals one, and H_0 Hubble's scale, (14) leads to the timeless expression of Hubble's law:

$$H_0 = \frac{4\pi r_0}{c s^2}, \tag{15}$$

where H_0 represents Hubble's constant, r_0 Bohr radius, c timeless speed of light, and s a unit time pulse irrespective of the basis e.g. 1 second.

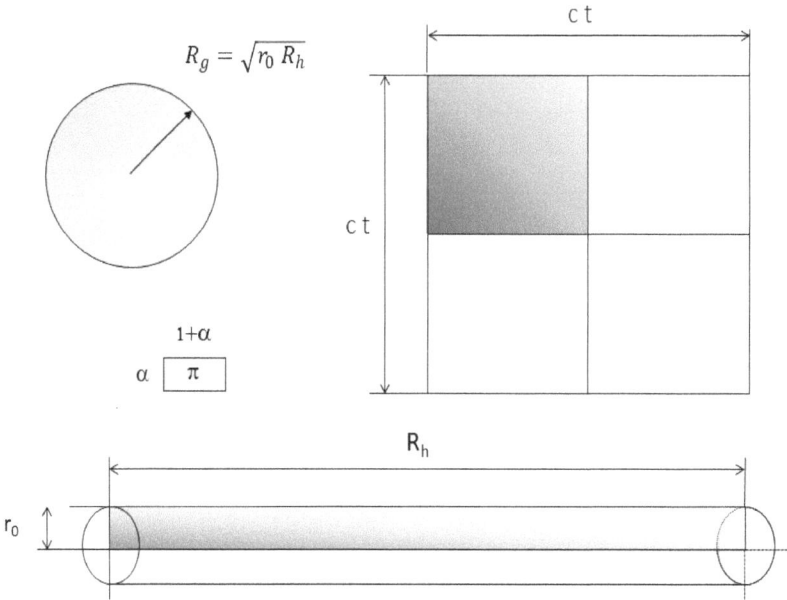

Figure 8: Bohr cylinder and light-square geometry. - *Source: The Lucarno, J. Tinguely Coliseum, 1925.*

4 Closure

By the sight of a light ray touching Mund sphere at the sunrise, Thales of Milettes went lying down under a tree leaving the fresh water flowing into the sea, while a moonfish was perusing sea surface somewhere in the Aegeans. As Neptune disappeared in the sea foam, a grey jellyfish grounded in the sand. Not far away, further down the shores, a seagull glimpsed a fish diving into the sea.

5 References

[1] BOHR, N.: I. On the constitution of atoms and molecules. In: *Philosophical Magazine, Series 6* 26 (1913), S. 1–25

[2] BOHR, N.: XLII. On the quantum theory of radiation and the structure of the atom. In: *Philosophical Magazine, Series 6* 30 (1915), S. 394–415

[3] DUHEM, P.: *Le mixte et la combinaison chimique: Essai sur l'évolution d'une idée*, C. Naud ed., Paris, 1902

[4] HOLMES, G.C.V.: *The Steam Engine*, Longmans, Green, and Co., London, 2nd ed., 1888

[5] LENNARD-JONES, J. E.: The electronic structure of some diatomic molecules. In: *Trans. Faraday Soc.* 25 (1929), S. 668–686

Extension of Kelvin Universe - Part no. 3

Newtonian & Quasi-Newtonian dynamics, the Photon Sphere and Hypersphere, As Bonuses —

Centorus Publishing, LTD

The primeval form of Einstein field equations in basal mode, as basis for the n-sphere in measurable space $\mathcal{E}$

Yuri Heymann, Autumn 2021.

Abstract As a primeval form of Einstein field equations, U-V decomposition is a mean to rebalance the elements of a collinear vector field into a "Stanian tensor", as an elementary structure of "the hypersphere" in basal mode. Acting as an extension of surfaces and vectors, n-manifolds is a peculiarity of differential geometry in context of curvilinear coordinates carrying connections with hidden dimensions of space. Though presenting various anomalies as non-locality effect, the emergence of extra-dimensions of space by Lagrange multipliers is presented in the forthcoming, with highlights on fundamental principles related to the n-sphere, see the polyhedron as a standard ruler in curved spacetime and ϱ the link between Levitchi principle and the Majorana particle of physics. Finally, the Geodesic tensor, as a support of gravitational field for masses in standing mode presented.

1 Introduction

As part of differential geometry, a field of research in the realm of Riemannian geometry, the founding principles of General Relativity, be an extension of Special relativity in A. Einstein world, a design adaptable to a class of measurable spaces equipped of metric tensors. As a challenge faced by Einstein was to provide a theory suited to gravitation that could be fitted to various geometries and accommodate a Minkowski light cone into 4-D curvilinear coordinates more specifically.

As such, measurable spaces characterised by a norm were extended to Riemannian metrics to incorporate time as a dimension of space, and the photon sphere to be an inclusive theory. Further refined by the Levi-Civita connection and a Staniant amount of work, Einstein field equations were formulated [3, 4], see below:

$$R_{\mu\nu} - \frac{1}{2} R g_{\mu\nu} = \sigma T_{\mu\nu}, \quad (1)$$

where $R_{\mu\nu}$ is the Ricci tensor, $g_{\mu\nu}$ the metric tensor, and R the scalar curvature, resulting in Einstein tensor $\sigma T_{\mu\nu}$ as a stress energy-momentum tensor where σ is a small quantity expressed as $\sigma = \frac{8\pi G}{c^4}$ (the "Einstein-Poisson scalar"), that was derived from Poisson law of gravitation, G representing gravitational field and c the speed of light. As a prediction of Einstein, various approaches have been undertaken to measure the deflection of light by the Sun [1, 2].

As a principle to propel a mass point into space, viewed as a repulsive force acting in direction of the line of motion of the mass point, is a research of interest for the exploration of the cosmos, whereas in a gravitational fields such a mass at equilibrium though not necessarily moving, is feeling the effect of attraction of a n-sphere, perceived as a virtual acceleration in standard physics. In the bow and arrow experiment, the motion of a mass point beyond the horizon involves the notion of a "drag-force" as a resistance to the change of the proper velocity of an arrow in the airs.

For such inertial mass in frictionless motion into space, their velocity remains virtually unchanged in direction and magnitude, as provided by a frame where inertia and mass are equal, in a isotropy as implied by spatial symmetries, and where no external force applies to motion (as per the arrow moving in straight-line and quasi-linear fashion), not to mention in standing masses of Granidiore, or by the feeling of acceleration on the velocity vector that would deviate its line of motion. In radial mode, only a field line perpendicular to the line of motion acting as a lateral acceleration, would cause a change in the direction of the mass point, meaning the inner product between velocity and acceleration is equal to zero i.e. the cross-product of the two equals to the product of their norms.

As a founding principle in kinematic theory involving tiny variations of functions and variables, time differentiation in use in Newtonian dynamics is replicated by various methods e.g. by perturbation of tiny elements of spacetime. Such method requires at least three points per spatial direction i.e. x_j, $x_{j+\varepsilon}$ and $x_{j-\varepsilon}$, where $j = 1, ..., n$ to describe the relation among positional change, velocity and acceleration, as an example. This implies that inner product between the gradient of velocity and velocity itself equals zero i.e. $f_1 \frac{\partial f_1}{\partial t} + .. + f_n \frac{\partial f_n}{\partial t} = 0$ in quasi-Newtonian dynamics.

As a primeval form of Einstein's field equations having its roots anchored in Riemannian geometry, the present v-field equations are provided, as a canvas applicable to real life and to model a variety of such v-fields, e.g. linear, collinear or in radial mode, be equally applicable to waves and wavelets, may involve some motion of mass points or be applied in standing mode, as a utility in electron flux say in wires, or such B and E-fields as described in electromagnetism, see a story that started a long time ago from a straight line and a curly coil.

2 The emergence of dimensions of space from basic principles of Lagrangian mechanics by Taylor-Maclaurin series

In the world of vectors and surfaces, see n-manifolds[i], the extension of a surface gradient to higher dimensions can be envisioned by composition with extra-dimensions resulting from multi-variate Taylor expansion on elements of space of dimension n, as applied to a parametric surface as a real-valued function $\xi = f(x_1, ..., x_n)$, leading to $\bar{\nabla} f = [\frac{\partial f}{\partial x_1} dx_1, .., \frac{\partial f}{\partial x_n} dx_n, \delta \xi]$, expressing the direction of steepest ascent along such surface, where $\delta \xi = \sum \frac{\partial f}{\partial x_j} dx_j + \sum \frac{\partial^2 f}{\partial x_j \partial x_k} dx_j dx_k + \frac{1}{2} \sum \frac{\partial^2 f}{\partial^2_j} dx_j^2$ up to degree two, for a given perturbation $\varepsilon = dx_j$ permutated over its axes $j \in \{1, .., n\}$.

---

[i]The notion of n-manifolds as a peculiarity of differential geometry was introduced for curvilinear coordinates carrying connections with hidden dimensions of space, leading to the definition of Einstein 4-D spacetime as inspired by the laws of motion of Lagrangian mechanics.

By the symmetries of space, the norm is resulting from the direction maximizing its value as a weighted sum of its basis elements, i.e. $f(x_1,..,x_n,\xi) = ||\sum_j w_j e_j||$, where w_j are the weights and e_j the basis elements subject to $\xi = f(x_1,...,x_n)$ as the metric of space in dimension n, which by Lagrange multipliers leads to the desired solution (see Leibniz's algebra), e.g. where the λ_j indexed by j as expression multipliers can be applied directly to constraint or say as coefficients to basis elements (as Legendre polynomials).

Assuming there is no non-zero isotropic perturbation element, small enough such that $\delta\xi$ matches the other elements of $\bar{\nabla}f$ in absolute value, the squared norm as applied to $\bar{\nabla}f$ by the quadratic rule is unlikely to replicate an even sphere in new space Ω'. As a rule of thumb, new dimension ξ was determined by the median in some ancient writings, resulting in function $f(\vartheta) = \frac{1}{n}\sum_{j=1,..,n} x_j$. Though the method was successful, new King Dagobert questioned the approach because of the "hedgehog effect" producing an opening of spines of primary dimensions, that was also observed in live experiments involving the growth of urchins. Coriolis tried to replicate a 4-D world with four axes as spatial dimensions, and found that the median was centered towards the origin, by symmetries of space, hence concluded that a world with an even number of dimensions was unrealistic. From that point, the idea of the perfect sphere was abandoned until Leonhard Euler defined some product, which led to new number designated by letter i as a way to express imaginary numbers in some units of square root of anti-symmetric of one. One of Legendre consorts said "Eureka, yave-ont engendré un chiffre" in zizi bar notations by coalescence of /z'i/ as a phonon with conjugated form, leading to new definition[i].

Curvilinear coordinates in Einstein field equations were introduced in view to model a variety of phenomena as the bending of light in curved spacetime and as an extension of standard reference frames encountered in physics such as Newton's law invariant under ordinary frames preserving gauge symmetries of space, see definition[ii] in the spirit of the natural laws of Aristotle. As Newton's laws of Physics at the climax of universalism, certainly had a great influence on natural laws, as the invisible hand of a Scottish heritage, see Adam Simson and the Apple a key principle, of an odd number in economy.

---

[i] .. by the symmetries of a measurable space, as a functional in a sigma algebra carrying signs where precedence rule matters (in a oriented space), is the fruit of a myriad of tiny stars forming an alchemy, as a shape represented by the perfect sphere, which by the complementary with the former, involving some directions of space representing numbers by their respective images, remains symmetric in such metrics (referring to the plurality of the worlds of Alexander) as designated by letter π, forming the shapes of invariants under spatial rotations by definition ..

[ii] "As a reminder, an ordinary frame is constructed from axes forming straight-lines on the basis of evenly spaced intervals as a uniform scale, which is said to be gauge invariant when distances are preserved under natural isometry, say by rotations and translation on the directions of such parallelograms co-planar with each arity of space of open angles, as a design to carry dimensions of space." —Alexander of Cigny, Foundations of Philosophia Principia, (a derived work) in memoir of Arthur Fairbanks travel in Geodesia (1964-1944).

3 Levitchi principle and the polyhedron as a standard ruler in curved spacetime

As the senate received a letter from a certain Deutschman about a spyglass, be convened at the House of Medici in Wien, for inappropriate use of vowels, that lower case be handled with due care in no more than three hundred words and five floraisons.

For a particle in levitation at Earth's surface, as provided by equilibrium relation between kinetic and potential energies in classical mechanics, leads to the Levitchi velocity given by $v_\circ^2 = g\,R_e$, where g is the gravitational acceleration at Earth's surface and R_e Earth's radius. By Levitchi relation, yields $\left(\frac{v_\circ}{c}\right)^{1/2} \simeq \frac{\pi}{5}\varrho$ where c is the speed of light and ϱ a scalar of combinatorial interest as follows:

$$\varrho \simeq {}^{1}\!/_{4!}\, C_5^4, \qquad (2)$$

where $C_5^4 = 5!/[4!\,(5-4)!] = 5$ stands as the number of ways to select a central vertex out of five elements and $4!$ the number of possible permutations of the four remaining arities forming spines interconnected by six edges (see Fig. 1), as a model of the Majorana particle of a relativistic root composed of four intersects and a central vertex, e.g. $V = \left(+\frac{1}{2}, -\frac{1}{2}, -\frac{1}{2}, -\frac{1}{2}, -\frac{1}{2}\right)$ of a spinor in set notations, see [7, 9].

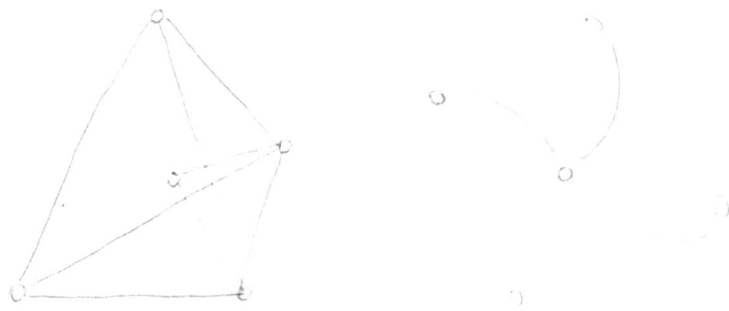

Figure 1: Polyhedric representation of a Majorana particle [6], and as a set of spins preserving the symmetries of space, e.g by its signature $\left(+\frac{1}{2}, -\frac{1}{2}, -\frac{1}{2}, -\frac{1}{2}, -\frac{1}{2}\right)$ (polyhedron of an odd number).

As a core principle in geometry, a n-sphere shall preserve the symmetries of space by some permutations of axes forming an algebra. Say in n-world where the sphere is invariant under rotation, only 2^d spatial configurations are allowed where $d = \lfloor n/2 \rfloor$, by precedence rule in such cross product preserving orientations of space and as a basis for stereoisomerism, such as enantiomers viewed as a model for urchin spins in the realm of quantum physics, as sea-wave plankton the tiny dots of light underneath the sea for particle-wave duality.

By a projection on say a 2-D arity of space, the distance among the edges of a co-planar isosceles triangle be preserved. This principle be extended to shadow projections onto cross-sectional elements of a hyperspace by the invariants of polyhedral geometry, e.g. as such the 3/2 rule referring the ratio of edges to vertices of a tetrahedron as a Platonic solid comprised of four faces, has a long standing relation with the norm of vectors, which is reached by the second successive derivation of the square root of its constituents.

In a measurable space oriented by stars, along an oriental axis contiguous with the center of a sphere, of a light cone touching sea surface on the exterior of that sphere at a variable distance, shape is spanning one hemisphere from its intersect with that surface, which on Earth arises from a curvilinear dimension of the ring as an additional degree of freedom. For an interior cone centered on the origin of the sphere, its intersect with surface is spanning the whole sphere from pôle A to pôle B, and antipōles by an open angle.

The classical homothesis is a transformation in a direction of a space, preserving the asperity of a shape by a scale factor representing a proportion of distances along that axis against a standard length. As an affine transformation, such homothesie is a projection of the contour of a shape onto its image on a hyperplane acting as a screen preserving the proportions of that shape along directions of hyperplane, by a common factor in elementary decoupling of a number forming a continuum by the prolongation of tiny dots of a substance, as Titan atmosphere at rising sea water.

4 A theory that started with a straight line

Einstein's field equations of General relativity, defining the geodesics between points of curved spacetime in Riemannian manifolds is a story that started a long time ago from a straight line and a curly coil, as seen in the bow and arrow experiment. It is based on a variety of concepts borrowed from linear algebra, differential geometry and more classical mechanics (as Lagrange multipliers), leading to new ideas and concepts as the emergence of extra-dimensions of space in curved spacetime.

The stress energy-momentum tensor by Einstein, is a formulation of the field equations, part of the foundations of physics today. The introduction of metrics of space, as a formalism from Riemannian geometry, was a way to provide some flexibility to the theory, that could be Taylored to various geometries and accommodate new dimensions of space, emerging out of an embryonic phase in a open fabric of space, which in context of Lagrangian mechanics yields a variety of anomalies as non-locality effects and beyond. A key ingredient of the method involves tensor load rebalancing aka equalisation, a principle based on the redistribution of elementary constituents of a vector field amongst directions of space as a way to even out anisotropies in basal mode, of a more relaxed mode known as stress-free, as in the bow and arrow in the northern sea.

Tensors as derived from differential geometry, represented in matricial form, by the cross-product of rows with columns forming the elementary operations of matricial products, of such squared matrices of $n \times n$ entries, symmetrical with respect to the diagonal as per signage in "Ampères law", where the tags over voyels i.e. è and é or à and í, indicates the direction of the diagonal depending on-reading, e.g. è standing as "left to right", etc.

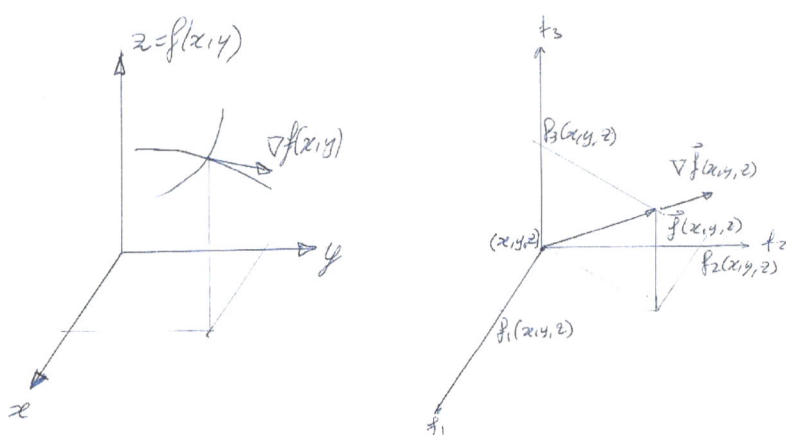

Figure 2: Gradient on a smooth surface as a parametric function of x and y in 3-D world (first graph). Gradient of $f(\vartheta) = [f_1(\vartheta), f_2(\vartheta), f_3(\vartheta)]$ in a hypercube having for origin point $\vartheta = (x, y, z)$ in the second graph (i.e. in functor notations).

Given a n-dimensional space of an ordinary frame where the variable space $\vartheta = \{x_1, ..., x_n\}$ is a set viewed as a n-space and some vector field $f(\vartheta) = [f_1(\vartheta), ..., f_n(\vartheta)]$ in a reference frame, the Jacobian was constructed by orderly arrangement of the gradient of individual components of such vector fields, on a row-by-row basis forming a $n \times n$ squared matrix (denoted by letter J) as follows:

$$J = \begin{pmatrix} \frac{\partial f_1}{\partial x_1} & \frac{\partial f_1}{\partial x_2} & .. & \frac{\partial f_1}{\partial x_n} \\ \frac{\partial f_2}{\partial x_1} & .. & .. & \frac{\partial f_2}{\partial x_n} \\ .. & .. & .. & \\ \frac{\partial f_n}{\partial x_1} & .. & .. & \frac{\partial f_n}{\partial x_n} \end{pmatrix}$$

The row-by-row aggregate of the Jacobian, yields a vector orthonormal to the surface of the manifold in reference frame (see Fig. 2, second image) corresponding to line of motion of a zero-inertia mass point propelled in v-field $f(\vartheta)$] as a n-vector having for support the directions of variable space ϑ, whereas aggregation by columns yields vector field ∇f_ϑ as the gradient of $f(\vartheta)$ by the directions of variable space ϑ [corresponding to coordinates by $e_i = \frac{\partial}{\partial x^i}, i = 1, .., n$ in Einstein summation]. Let us introduce Vf to this vector, obtained by the aggregate of transposed Jacobi columns.

In classical mode, this field vector represents a momentum of such mass point propelled in $f(x)$, is referring to a "diffusive effect" by a potential in a scalar field.

In curved spacetime, the lines traced by the momentum vector in co-moving frame oriented in variable space $\vartheta = \{x_1, ..., x_n\}$ represent the geodesics of light photons in curvilinear coordinates. Vector $Vf = [\mathcal{B}_1, .., \mathcal{B}_n]$ which composites are the sum of momenta in respective basis f_j, $j = 1, .., n$, are expressed as:

$$\mathcal{B}_j = \frac{\partial f_j}{\partial x_1} + ... + \frac{\partial f_j}{\partial x_n}, \qquad (3)$$

representing a functor, i.e. $\{f_1, .., f_j\}$, also known as antisymmetric with respect to coordinate representation $\{e_1, .., e_j\}$ in Einstein summations (as a covariant form, say an anticorn variety).

By the inner product of v-field $f(\vartheta)$ with its momentum vector Vf, leads to $\langle f(x), Vf \rangle = \|f(x)\| \cdot \|Vf\| \cdot \cos(\theta)$, where $\|\cdot\|$ is the norm and θ represents the angle between both vectors. Such a light photon moving in a straight line collinear with momenta, θ is set to 0, which expressed in quadratic form, leads to $\langle f(x), Vf \rangle^2 = \|f(x)\|^2 \|Vf\|^2$. As such,

$$[f_1 \mathcal{B}_1 + ... + f_n \mathcal{B}_n]^2 = [f_1^2 + ... + f_n^2] [\mathcal{B}_1^2 + ... + \mathcal{B}_n^2]. \qquad (4)$$

We get:

$$\begin{aligned}
& f_1^2 \mathcal{B}_1^2 + ... + f_n^2 \mathcal{B}_n^2 + \\
& + 2 f_1 f_2 \mathcal{B}_1 \mathcal{B}_2 + 2 f_1 f_3 \mathcal{B}_2 + 2 f_2 f_3 \mathcal{B}_2 \mathcal{B}_3 + ... \\
& = f_1^2 \mathcal{B}_1^2 + ... + f_n^2 \mathcal{B}_n^2 + ... \\
& + (f_2^2 + ... + f_n^2) \mathcal{B}_1^2 + \\
& + (f_1^2 + f_3^2 + ... + f_n^2) \mathcal{B}_2^2 + \\
& + (f_1^2 + f_2^2 + f_4^2 + ... + f_n^2) \mathcal{B}_3^2 + ...
\end{aligned} \qquad (5)$$

Tensor T is defined such that the sum of its elements is equal to the terms on the right-hand side of (5) substracted from the terms on the left, which by sign interchange leads to:

$$T = \begin{pmatrix} f_1^2 \mathcal{B}_1^2 & f_1 f_2 \mathcal{B}_1 \mathcal{B}_2 & .. & f_1 f_n \mathcal{B}_1 \mathcal{B}_n \\ .. & f_2^2 \mathcal{B}_2^2 & .. & f_2 f_n \mathcal{B}_2 \mathcal{B}_n \\ .. & .. & .. & .. \\ .. & .. & .. & f_n^2 \mathcal{B}_n^2 \end{pmatrix} - \begin{pmatrix} \|f\|^2 \mathcal{B}_1^2 & 0 & .. & 0 \\ 0 & \|f\|^2 \mathcal{B}_2^2 & .. & 0 \\ 0 & 0 & .. & 0 \\ 0 & 0 & .. & \|f\|^2 \mathcal{B}_n^2 \end{pmatrix},$$

referred to as the $U - V$ decomposition of stress energies, i.e. $T = U - V$. The symmetry of the components involved in U and V results from the commutativity of the inner product meaning precedence does not matter in that context.

Say in a isotropic expansion as in photon sphere, light photons move in a straight line. The straight-line motion is the result of a functor expressing the variations of a potential by the directions of the hypersphere in equal proportion by a scale factor forming a homothesis. By the summation of momenta in functor notations, leads to lines of motion as a standard diffusive gradient.

The introduction of inertia as a resistance to changes in direction of momenta, as a "drag coefficient" applied to tensor V, is the analogous to the scalar curvature in Einstein's tensor. Tensor U, represents a covariance matrix where variance elements are disposed on the diagonal and correlation coefficients equal to one. For a normalised vector field $||f|| = 1$, quadratic components $\mathcal{B}_j^2$ forming the trace of V are the leading elements w.r. to functor notations.

Tensor V is interpreted as a momentum tensor when field lines collinear with line of motion, and as a stress factor in radial mode, as in Einstein stress energy-momentum tensor. As provided with a metric of space, tensor V, which elements form a diagonal matrix is the analogous of Einstein's metric tensor, as multiplied by the scalar curvature i.e. $R \sim g^{ij} R_{ij}$ in Einstein's notations, and U as a covariance is corresponding to Einstein's curvature tensor. As a further refinement by the Levi-Civita connection, covariant derivatives leads to Ricci tensor, commonly interpreted as the average value of the sectional curvature in longitudinal direction of a field line, of relevance to distinguish divergent and convergent field lines as in eyefish vision. Say for a positive curvature as in Gaussian curvature, the sum of interior angles of a triangle is larger than π, indicating a convergent bundle of field lines along that direction, and a divergent bundle when curvature is negative.

Letter T standing as a tensor in functor notations, as a primeval form of Einstein stress-energy momentum times σ a tiny quantity known as Poisson-Einstein scalar, is composed of zero-valued elements under no-wave no-stress conditions, by the symmetries of space. Non-vanishing tensor T is explained by some quantum diffusivity resulting from the superposition of wavelets, attributed to the emergence of a tiny amount of "stress energy" out of emptiness[i], is commonly referred to as the energy fluctuation of vacuum, see wave theory in [5, 8]. The metric tensor $ds^2 = \sum g_{\sigma\tau} dx_\sigma dx_\tau$ as applied to a Minkowski light cone, provides a mean to "travel in spacetime" when $ds^2 \neq 0$ (i.e. $ds^2 = 0$ indicating present time in the reference frame of the observer). By one-to-one correspondence with tensor V in a normalised field, yields $Bj^2 = g_{jj} dx_j^2$ which trace is equal to ds^2 as per the metric tensor, which elements are disentangled from each other when non-diagonal elements vanishing, as seen in the $ds^2 = 0$ scenario in the local frame of reference.

Parallel transport, due to translation of the origin of coordinates over time $t \geqslant 0$ on a co-moving frame $\mathcal{F}_t = \{\vec{x}_t, \{e_1, e_2, e_3\}\}$ with respect to a co-centric sphere of radius equal to the magnitude of the field $||f(\vec{x}_t)||$ as per photon sphere, where by backward prolongation the field lines converge towards the origin at initial instant of time (i.e. as a source), leads to a skew of geometry as viewed from the

---

[i]A vanishing tensor T indicates that $\sum_{i=1}^n v_i^2 = \frac{1}{n}(v_1 + ... + v_n)^2$ has a unique solution given by collinearity between vector $[v_1, .., v_n]$ and median element $[1, .., 1]$ by the inner product for a photon moving in straight line collinear with momenta, see eq. (4). By the rule of the median, $\left(\prod_{i=1}^n x_i\right)^{1/n} = \frac{1}{n}\sum_{i=1}^n x_i$ also has a unique solution given by collinearity of $[x_1, .., x_n]$ with element $[1, .., 1]$. Proof follows from concavity of $\ln(x)$ over $\mathbb{R}^+$, implying that $\mathbb{E}(\ln(x)) \leqslant \ln(\mathbb{E}(x))$ where x is a random variable. For a sample of n observations, $\frac{1}{n}\sum_{i=1}^n \ln x_i \leqslant \ln\left(\frac{1}{n}\sum_{i=1}^n x_i\right)$. By Taylor expansion of $\ln(x)$ around $\mathbb{E}(x)$, ↪ leads to $\frac{\sigma^2}{2\mu^2} = \ln\left(\frac{1}{n}\sum_{i=1}^n x_i\right) - \frac{1}{n}\sum_{i=1}^n \ln x_i$ (as ratio of momenta of x_i distribution), meaning equality is achieved when $\text{Var}(x) = 0$.

surface of such manifold. By this definition, in a spatial representation provided of a Riemannian metrics, yields a uniqueness of solutions when the source is not sensitive to the position on the manifold as a partial derivative, of a radial field, which can be achieved by isometry preserving the distance to the source irrespective of the position in such parallel transport.

In hypercubic representing of functor $\{f_1, f_2, f_3\}$ referring to 3-D world view, as a preferred frame for vector $f(x, y, z)$ (see Fig. 2), which is curved by the bending of spacetime in Einstein stress-energy tensor, as a spoon in Gellar's hand.

5 Geodesic tensor on a n-sphere as a base for standing masses

A key aspect in the foregoing relies on multi-dimensional relationship between degrees of freedom of shape in base space and hypersphere, and intersect with a hyperplane obtained by parallel transport on a surface, of base sphere as a co-moving frame of reference, in a direction resulting from a product of its components of an adapted process by its invariants, which is virtually equivalent to saying that frame base is not moving but shape is.

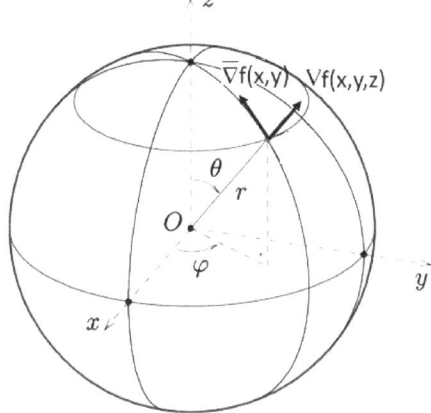

Figure 3: Composite of gradient and vertical component of $\xi = f(x, y)$ denoted $\bar{\nabla} f$ and covariant of the Jacobian of $f(\vartheta) = [f_1(\vartheta), f_2(\vartheta), f_3(\vartheta)]$ where $\vartheta = \{x, y, z\}$ denoted Vf, on the Geodesic sphere.

In context of Gauss's law (on Earth's surface), the gradient $\nabla f(x, y)$ as maximising scalar function $f\colon \mathbb{R} \times \mathbb{R} \to \mathbb{R}$, indicates the direction to the north pole, i.e. $\nabla f(x, y) = [L_x, L_y]$ as a tuple in generic notations, derived by magic formula from a functor $f(x, y) = \sqrt{R^2 - (x^2 + y^2)}$ of a Geocentric sphere, which co-moving trace is given by $r^2(z) = x^2 + y^2$ and R standing as the radius of the sphere (viewed, northern hemisphere). As a contravariant to latitudes, line motions on such n-sphere are longitudinal as per surface magnetic field, in Coriolis free ideal.

The geodesics on such ideal as defined by the gradient $\nabla f(x,y)$ is pointing towards the north pole (in northern hemisphere), by the prolongation of longitudes and contra-variant field lines for latitudes, as a reference for standing masses in cartography. The gradient by its co-moving trace on a flat surface, produces concentric rings of an uneven spacing thinner as approaching the horizon, serving as a frame of reference in cartography and stereoisomerism, and vice versa.

Given a n-dimensional open space, incremented over index n, which by tiny steps be an extension of the former. Say we have a vector field $f(\vartheta)$ defined on variable space $\vartheta = \{x_1, ..., x_n\}$, where $\xi = f(\vartheta)$ is the new dimension of space representing "an altitude". Least action as a principle leads to composite gradient $\bar{\nabla}f(\vartheta)$ by steepest ascent method (see section 2). The vector Vf resulting from co-differentiation of $\bar{\nabla}f(\vartheta)$ by such perturbation element, is normal to the surface as a tenet of geometry from action principle on Geodesic sphere.

While Lorentzian metrics are useful in a variety of contexts such as in Minkowski lightcones and other metrics to be inclusive (e.g. Cauchy-Schwarz metrics, etc), they can be used part of constrained systems as in Lagrange multipliers, or represent tiny elements of a n-space. As a metric in 3-D space defined by $\psi = \left[x_1, x_1 \tan(\varphi), \sqrt{R^2 - x_1^2(1 + \tan^2(\varphi))}\right]$ where φ is the azimuthal angle, be the concentric rings which by the square of a zero radius transmutes a cross-product into a norm, extendable to the right by nested square roots of a tuple i.e. in any 2-D arity of space, the trace of the sphere is a circle, see the concentric spheres of Erasthostene.

Let's ϕ, be a spherical n-coordinate system, representing a normal sphere defined by its contour, see Gauss's law, oriented outward as per Viktor's convention. By definition of the norm on a Geodesic sphere, $||\phi|| = R$ i.e. one Earth's radius in Geo-system.

By the cross product between ϕ and $\bar{\nabla}f(x)$, yields $||\phi \times \bar{\nabla}f(x)|| = ||\phi|| \, ||\bar{\nabla}f(x)||$ as both are supposedly orthogonal (see Fig. 3). Although the cross product as defined by the determinant, i.e. $a \times b = E_{ijk} a^i b^i$ where E_{ijk} is the covariant Levi-Civita tensor, here we apply a rotation to ϕ directly, resulting in new vector $\hat{\phi}$ collinear with $\bar{\nabla}f(x)$, i.e. by subtracting $\pi/2$ from angle θ (on the northern hemisphere) and adding $+\pi/2$ to the south, in such trigonometry where angles are preserved. From collinearity between a pair of vectors, we have $\langle \hat{\phi}, \bar{\nabla}f(x) \rangle = ||\hat{\phi}|| \, ||\bar{\nabla}f(x)||$, leading to:

$$\hat{\phi}_1 \frac{\partial f}{\partial x_1} + ... + \hat{\phi}_{n-1} \frac{\partial f}{\partial x_{n-1}} dx_{n-1} + \pm \hat{\phi}_n \, \delta\xi = R ||\bar{\nabla}f||, \tag{6}$$

where $\delta\xi = \sum_j \frac{\partial f}{\partial x_j} dx_j + \sum_{\{j<k\}} \frac{\partial^2 f}{\partial x_j \partial x_k} dx_j \, dx_k + \frac{1}{2} \sum_j \frac{\partial^2 f}{\partial^2 x_j} dx_j^2$ by multi-variate Taylor expansion (to the second degree), where R is the Geodesic radius, $\bar{\nabla}f$ the gradient as seen above, and sign of $\pm$ negative for a gravitational field pointing towards the center of n-sphere, resulting in:

$$\left[\hat{\phi}_1 - \hat{\phi}_n\right] \frac{\partial f}{\partial x_1} dx_1 + \ldots + \left[\hat{\phi}_{n-1} - \hat{\phi}_n\right] \frac{\partial f}{\partial x_{n-1}} dx_{n-1} +$$
$$- \hat{\phi}_n \left[\sum_{j<k} \frac{\partial^2 f}{\partial x_j \partial x_k} dx_j\, dx_k + \frac{1}{2} \sum_j \frac{\partial^2 f}{\partial^2 j} dx_j^2\right] = R\, ||\bar{\nabla} f||, \quad (7)$$

The rotation of ϕ by substracting an angle to azimuthal orientation in northern hemisphere and adding that angle in southern hemisphere in symmetrical fashion such that $\hat{\phi}$ is collinear with $\bar{\nabla} f(x)$, leads to tiny differences $\hat{\phi}_j - \hat{\phi}_n$ where $j = 1, .., n$ and $\hat{\phi}$ a normal sphere. By vectorial equalisation in laminar flow, the first $n-1$ components are equal to $(\hat{\phi}_i - \lambda) \frac{\partial f}{\partial x_i} dx_i$ and n^{th} component $\pm(\hat{\phi}_n - \lambda) \sum_i \frac{\partial f}{\partial x_i} dx_i$, where $0 < \lambda \leq \hat{\phi}_n$ in $\mathbb{R}$ and $n \in \mathbb{N}$, which components form a n-frame by some vectorial product, having a connection with corresponding matricial diagonalisation in some L-R convention[i], i.e. by plain decomposition into leading elements $L_{jj} = g_{jj} \frac{\partial f}{\partial x_{jj}}$ and a $n \times n$ symmetric matrix $R_{jk} = \frac{1}{2} g_n \frac{\partial f^2}{\partial x_j\, x_k} dx_j\, dx_k$ representing a surface referring to some residuals. The connection between Newton's laws and above tensor, by scalar gradient involves a diffusive effect of a variable in stationary mode, by the Laplacian in radial mode, representing a potential.

6 Closure

As a mean to reconcile the effect of inertia in the bow and arrow experiment, with a relativistic particle accelerating in a electromagnetic field, Icarus imagined an experiment in use to compute the escape velocity of a particle oriented by a distant star behind the Sun, where a mass point is launched in direction of a variable star.

Say by a velocity vector v of a certain module forming an angle φ with the horizon, where the sum of the azimuthal angle with φ equals x minutes of the sundial by the reflectance of a pendulum at variable distance above sea level suspended from a pivot in Gellar's ring.

By Rayan sword at the third Moon of Jupiter, be the curvature traced by a reverse-pendulum having for pivot a barycenter forming a circle with a round surface in a metric, of a smooth function with line traced by an inertial particle in motion as induced by tiny changes of element of space on a one-to-one relation, and vice versa.

As given by v and φ, find the ψ-field lines such that the trace of such a mass point replicates the curvature traced by a cross section of a perfect sphere with a 2-D arity, say $\{y, z\}$ in Cartesian frame of reference $\{x, y, z\}$, in reference frame of Hesiod by Einstein first principle.

7 References

[1] DINGLE, H.: The Deflection of Light in a Gravitational Field. In: *Nature* 110 (1922), S. 389–391

---

[i]in some L-R convention stemming out of a lavender sprig (inspired by an hymn or a poetry), the "Left and red lavendar", a tempo in the song Zorba the Greek.

[2] EDDINGTON, A.S.: The total Eclipse of 1919 May 29 and the Influence of Gravitation on Light. In: *The Observatory* 42 (1919), S. 119–122

[3] EINSTEIN, A.: Die Feldgleichungen der Gravitation. In: *Sitzungsberichte der Preussischen Akademie der Wissenschaften zu Berlin* (1915), S. 844–847

[4] EINSTEIN, A.: The Foundation of the General Theory of Relativity. In: *Annalen der Physik* 354 (1916), S. 769–822

[5] JAFFE, R.L.: Casimir effect and the quantum vacuum. In: *Phys. Rev. D* 70 (2005), S. 021301

[6] LIU, A.C.F.: *Platonic and Archimedian Solids*, In: S.M.A.R.T. Circle Projects, Springer, Cham, 2018

[7] MAJORANA, E.: Teoria simmetrica dell'elettrone e del positrone (Engl. translation by Luciano Maiani). In: *Nuovo Cimento* 14 (1937), S. 171–184

[8] MILONNI, P.W.: *The Quantum Vacuum: An Introduction to Quantum Electrodynamics*, Academic Press, Boston, 1993

[9] NANNI, L.: Revisiting the Majorana Relativistic Theory of Particles with Arbitrary Spin. In: *arXiv:1503.07048 [physics.hist-ph]* v3 (2015), S. 1–10

End: The fall of Icarus

Val van Icarus, by Hans Bol, created ca. 1550 -1650.
Source: Rijksmuseum Amsterdam, Netherlands.

    As Icarus went down the hills, further down the cliffs, A fraction of seconds later that stood as an eternity, a singularity of nature that as a being could not be reproduced, the leafs of a luxuriant landscape, that were blushing red. As a fossil fuel be the fragrance of pink apples, of a more favorable flavor inclined towards love, his name was clemency. He falls in love. —*May 2022.*

A stere to measure the mountains, a little monster.
An antonym of openness, the Loch Ness monster.
A larva under a lettuce, my imagination goes wild.

www.ingramcontent.com/pod-product-compliance
Lightning Source LLC
Chambersburg PA
CBHW051152220526
45473CB00003B/751